Energy Policy: Industry Perspectives

A Report to the Energy Policy Project of the Ford Foundation

Energy Policy: Industry Perspectives

John E. Gray

Ballinger Publishing Company • Cambridge, Mass.
A Subsidiary of J.B. Lippincott Company

Published in the United States of America by Ballinger Publishing Company, Cambridge, Mass.

First Printing, 1975

Library of Congress Catalog Card Number: 74-22454

International Standard Book Number: 0-88410-314-5 (H.B.)
0-88410-315-3 (Pbk)

Printed in the United States of America

Library of Congress Cataloging in Publication Data

Gray, John E
Energy policy.

Includes bibliographical references.
1. Power resources–United States. 2. Energy policy–United States. I. Title.
HD9502.U52G7 333.7 74-22454
ISBN 0-88410-314-5
ISBN 0-88410-315-3 pbk.

Table of Contents

Preface

The Energy Policy Project was initiated by the Ford Foundation in 1971 to explore alternative national energy policies. This book, *Energy Policy: Industry Perspectives,* is one of the series of studies commissioned by the Project. In this book the author has presented a composite picture of the energy industries' views on many of the policy issues that make up the energy policy debate. We believe this view from behind the corporate veil is important because national energy policy cannot be well fashioned without an understanding of attitudes and opinions of energy company executives.

It is our hope that each of these special reports will stimulate further thinking and questioning in the specific areas that it addresses. At the very most, however, each special report deals with only a part of the energy puzzle; our final report, *A Time to Choose,* attempts to integrate these parts into a comprehensible whole, setting forth the energy policy options available to the nation as we see them.

This book, like the others in the series, has been reviewed by scholars and experts in the field not otherwise associated with the Project in order to be sure that differing points of view were considered. With each book in the series, we offer reviewers the opportunity of having their comments published in an appendix, but none chose to do so with this volume.

Energy Policy: Industry Perspectives is the author's report to the Ford Foundation's Energy Policy Project and neither the Foundation, its Energy Policy Project or the Project's Advisory Board have assumed the role of passing judgement on its contents or conclusions. We express our views in the Project's final report, *A Time to Choose.*

S. David Freeman
Director
Energy Policy Project

Author's Note

This book is based on a study of influences on decision-making in the energy industry commissioned by the Ford Foundation's Office of Resources and Environment in September of 1972 as part of the Foundation's Energy Policy Project. The primary focus was on case studies involving the oil, electric utility, coal, gas and nuclear industries to determine the influences of the various interests that motivate management, and interaction between governmental policy, law and regulation, or the lack thereof, and industry decision-making, with emphasis on how a national energy policy could be fashioned so that private industry will have both the incentive and capability to carry it out. The study's conclusions on influences, observations on national energy policy, and representative material from the study report are included in the appendixes.

Acknowledgments

Companies cooperating in the research phase of the study included Commonwealth Edison, Consolidated Edison, Northeast Utilities, Carolina Power and Light, American Electric Power, TVA, Southern California Edison, Pacific Gas and Electric, Portland General Electric, City of Memphis Gas and Water, City of Jacksonville Electric Authority, Dallas Power & Light, Mobil, Exxon, Texaco, Shell, Arco, American Petrofina, Gulf Oil, Boston Gas, Pacific Lighting, Columbia Gas, Consolidation Coal, Westmoreland Coal, Peabody Coal, Eastern Coal, Amax, Westinghouse, Kerr-McGee, General Electric and Bechtel. First National City Bank and William D. Witter, Inc. also helped.

Industry associations contacted were the Edison Electric Institute, American Public Power Association, American Gas Association, American Petroleum Institute, Independent Petroleum Association of America, Atomic Industrial Forum, Independent Natural Gas Association and the National Coal Association.

Many thanks go to Sherman Knapp, Arleigh Burke, and Gene Zuckert for their initial encouragement and insights. Harris Ward, "Tav" Tavoulareas, John Cochran, Eli Goldston and Joe Rensch were most helpful in making themselves available for testing and refining the interactive (with industry) analytical methods used in the study. Don Crawford and Alex Radin were particularly helpful in providing insights regarding, and cooperation with, the private and public sectors of the electric utility industry. The same is true of Carl Bagge for the coal industry.

Bob Gray and Kathryn Dohet were most helpful in the preparation of the draft study report. Finally, John Adams Associates undertook the task of helping to prepare for publication a manuscript suitable for the diverse audience to which the Foundation's Energy Policy Project is addressing itself.

Energy Policy: Industry Perspectives

Chapter One

Introduction

The American energy industry is a complex web of men, resources, companies and laws that has evolved to meet the nation's demands. For more than 100 years, from the days of whale oil to the days of nuclear power, it has expanded and adapted to provide for the nation's industrial growth.

Traditionally, the mission of the U.S. energy industry has been viewed as one of meeting the demands of the United States (and in some instances other nations) for all forms of energy at the lowest price consistent with a reasonable profit. For the investor-owned gas and electric utility industries, this has been mandated by law. Individual industries and companies are expected to advance their own interests and to worry about the interests of others as they are required to or think they should. Decisions are made to this end within the constraints of, or with the encouragement of, public policy, law and regulation.

Until recently, decision-making within the energy industry was orderly and relatively easy, and the consuming public was, for the most part, content.

Traditional influences—supply and demand, service obligations, return on investment, "enlightened self-interest"—remain, but their impact, and the industry's ability to respond to them, has changed dramatically largely as a result of circumstances beyond the industry's control.

The U.S. fuel reserves base, particularly for the "Lower 48" states, has become inadequate to meet demand for oil and gas, while coal and nuclear power remain abundant—at least in potential. At the same time, it has become clear that other nations' decisions on the sale of oil to the United States will be based on their perception of their own self-interest, not America's.

Concurrent with this came the environmental and health and safety restrictions that constrain many traditional practices in the energy industry, and encourage the public to intervene in industry/government decision-making dealing with extent, character, location, and operation of industry facilities.

Price controls have further complicated matters. Gas and electric prices have traditionally been regulated, but Nixon administration economic policy imposed regulations, at least temporarily, on oil and coal as well.

Thus the industry today is laboring under an increasingly complicated set of exterior factors, particularly price and supply limitations. Government and industry have not yet found the mutual ground for decision-making that characterized their relationship into the 1960s.

The current situation is new to the public as well. Americans have been used to expecting their local electric and gas utilities and their neighborhood gasoline station to provide them with as much as they want whenever they want it and at a low price.

When one of these suppliers finds itself unable to live up to this image, as many have in the past few years, the result has often been suspicion and hostility. Utilities are accused of subverting regulatory agencies to allow excessive prices; oil companies are accused of holding back supplies to drive prices up; gas companies are accused of understating their reserves to gain rate increases, and many energy companies are accused of being indifferent to public health and safety and the environment.

These hostile reactions are quickly transformed into political pressure, and political pressure into price limitations or restrictive legislation—which can be ill-considered and make matters worse.

This adversary relationship between the energy industry and the public, which broke into the open on Capitol Hill during the Arab oil embargo—and which had been simmering in the background of many earlier disputes involving regulation of electric and gas utilities and the coal industry—is unlikely to benefit anyone. Restrictive legislation is likely to depress supply and to increase prices rather than solve the problem.

In the current atmosphere, some industry critics regard government takeover as the cure for the energy crisis. But government-run companies have the same problems in developing resources, providing supply and conforming with health and safety and environmental regulations. They must incur the same costs, and if these are not covered by revenues, they must be covered by the taxpayers. The problems and the costs remain.

The energy crisis, then, is a crisis of the national decision-making process as well. The public, the government and the industry are at a crossroads; fundamental choices must be made, and soon:

1. How much energy the nation needs vs. what the public will pay.
2. What the public will pay vs. its desire for a rapid environmental cleanup.

In a nation of cheap energy, lacking sensitivity to environmental degradation, choices were easy. Airplanes and high horse power cars were favored over trains, buses and subways; plate glass over insulated walls. High sulfur oil and coal were produced and used with emphasis on minimum cost, not minimum environmental impact. And the American lifestyle developed accordingly.

It will be a long time before the nation can again be assured of adequate energy supplies at stable prices. Increasingly, the nation will have to decide to what extent it is willing to pay more or to what extent it is willing to cut down.

Cutting down has serious implications for the American economy and lifestyle. The biggest energy users in the U.S. economy are industry and commerce, transportation, and domestic heating and cooling.

Most industry officials feel that adjustments need not be as severe as is often suggested, nor do they feel it is possible to presume in advance what the adjustments should be for the countless individuals and institutions involved. If prices are allowed to rise to keep pace with costs, private enterprise will have the incentive to press forward to meet demand, while at the same time the self-selection process that price provides in the free market will limit demand while shifting the heaviest burden to those who choose to use the most.

If the free market is to operate effectively, however, outside forces, including all pertinent law and regulation, must be based on an understanding of first, the realities of the supply problems to be faced in meeting national demand; second, the feasibility and cost of attaining environmental and other goals; and third, a realistic perception of what is involved in balancing energy supply and environmental goals.

In the case of environmental and health and safety restrictions which, except for availability of foreign oil and price regulation of utility companies, are the major exterior forces impinging on the industry today, realistic levels are those that are known to be achievable in a timely manner and at costs the nation can afford.

Most if not all companies in the industry today endorse the goal of a cleaner environment. But many also, at one time or another, find themselves confronted with an environmental requirement that cannot practically be met. Others have met delays so extensive they have added years, sometimes literally as much as a decade, to the process of planning and building new facilities.

In general terms, these problems face all areas of the industry–coal, oil, gas and electric utilities. In addition, there are more specialized problems facing each segment.

Coal, long the nation's predominant fuel, is just emerging from a period of eclipse. As demand and prices remained low during the 1950s and much of the 1960s, coal mines were closed and the miners dispersed. Technology, both in mining and coal use, lagged.

Then came the Arab oil embargo and the broader realization that coal is the only fossil fuel America has in sufficient quantities to allow real hope of independence from outside sources. In 1974, the coal industry was trying to expand its capacity, develop new uses and enlarge its share of the fuel market.

This intended renaissance is inhibited by many problems. Sulfur oxide restrictions, mine safety and strip mine regulations are issues that must be resolved if coal is to be used on a scale commensurate with its reserves. A stable market is also needed: coal-related regulations imposed on the electric utility industry, the primary customer for coal, must be both realistic and stable, so that the utilities can know that they are going to be able to continue to use coal and thus safely sign long-term contracts with suppliers. Such contracts, in turn, guarantee producers revenues, which give them the incentive to expand capacity.

Even when the Arab oil embargo struck, variances from federal and local air pollution controls granted to allow utilities to burn coal were only temporary, applying only to the period when imported oil was not available. With the return of oil, most of the power plants that had converted to coal began switching back. "My guys responded after the embargo," remarked Carl E. Bagge of the National Coal Association. "They increased production, bought new equipment, opened new mine faces. Now some of my guys are getting canceled."

By May, 1974, Congress was considering legislation to continue to permit the burning of coal, but the situation illustrates the problems of regulating an industry as complex as energy: coal companies cannot finance expansion without long-term contracts with utilities; utilities on the other hand are hesitant to sign such contracts for fear that coal companies will not be able to deliver or that environmental restrictions will again be imposed; finally, coal companies are hesitant to commit themselves to long-term contracts for fear that rising world prices will make the agreed-on price too low.

The oil industry confronts a swelling demand at the same time nations holding large reserves abroad are finding it desirable to limit production, nationalize U.S. oil companies' facilities, and escalate prices. Domestic U.S. oil wells have been at full production since 1970, making the nation increasingly dependent on imports—the Arab embargo demonstrated just how dependent.

The industry today is seeking new sources of supply within American control or in politically stable areas. The search is costly and, notably in the case of the North Slope of Alaska, production has been hampered by environmental concerns. Oil companies are confident they can overcome their present difficulties provided they are given sufficient economic incentive and access to resources under U.S. control—and if environmental regulations are applied in a manner that allows industry to develop resources and build facilities.

Natural gas, like oil, faces a problem of supply. The gas producing industry, operating under close government price control, has found domestic exploration and development less and less economical. New discoveries fell behind consumption in the latter part of the 1960s, and have not recovered.

New supply sources, such as importation of liquified natural gas or production of synthetics, have been hampered by government regulations and price structures.

Gas, which burns clean, is highly desirable environmentally. Also, burned directly for space heating or in industrial processes, it is economically desirable, especially at the regulated low prices at which it has been available. Consequently, demand gained steadily. The primary issue, the industry feels, is price—if it is allowed to rise, incentives will increase for exploration and development, and alternative processes, such as synthesis from coal or oil, will become economical.

The electric utility industry's problems are linked to those of its suppliers of generating equipment and fuel. Shortages and rapid price changes make it difficult for companies to plan new facilities and obtain rate adjustments in an orderly fashion.

In addition, environmental pressures are intense. New plants are often challenged in the regulatory agencies, in the courts, and even through demonstrations in the street. Frequently these challenges are successful, necessitating extensive changes and added costs.

Added costs are particularly significant to utilities because they are not free to raise prices. The agencies that regulate their rates often are slow to act and do not necessarily give the companies the rates they seek and need to provide reliable service now and in the future.

The electric utility industry needs help from the regulatory agencies not only in the area of obtaining adequate rates but also in obtaining sites, rights of way and licenses for construction and operation of nuclear power plants.

Today, many electric utility companies, because their profits are too low, are facing acute problems raising capital to build plants and maintain existing facilities. The problems have long-term implications for the utility industry. Growing demand, new technology, and high standards of service will require costly new plants. Companies unable to raise construction funds will suffer and their customers will suffer. Older, less efficient plants will remain on line longer; service breakdowns will be more numerous. The industry does not believe the public wants this, but also understands the public's desire for the lowest possible rates. What the public should understand, the industry feels, are the choices involved. Lower rates now put in jeopardy the availability of adequate electric supplies later.

The various segments of the energy industry, though they have individual problems and opportunities, operate in a single context. To whatever extent one fuel is used or not used, the others are affected. When environmental regulations led to a sudden massive shift from high-sulfur, eastern and midwestern coal to low-sulfur oil and low-sulfur western coal, shortages developed and prices rose.

Had the environmental restrictions been imposed more gradually and flexibly, adequate supplies of low-sulfur oil or processes for using coal in an environmentally acceptable manner or nuclear power might have developed in time to fill the gap. As it is, coal companies are struggling in a cloud of uncertainty to raise production, and utilities are struggling in the face of a very difficult supply and regulatory system to bring additional nuclear generators on line. Costs are rising and planning is difficult.

Following is a consideration of each of the four segments of the energy industry separately, an examination of the issues facing each, conclusions and recommendations.

Chapter Two

Oil

For most of its peacetime history, the U.S. oil industry has operated as a largely free enterprise, and the companies in the field are proud of their record of providing all the oil Americans have wanted and at a reasonable price.

Today the industry is in the initial stages of a profound change. Supplies are no longer readily available to provide for ever-increasing demand. While economic growth and environmental concerns are increasing the need for all petroleum products, known domestic supplies are being reduced, new domestic supplies are harder to find and less economical to produce, and foreign governments are less willing to sell their crude at or even near traditional price levels.

Since the mid-1960s, these factors have converged to threaten the industry's ability to provide adequate supplies at tolerable prices. The dramatic effects of the Arab oil embargo of 1973-74 have alarmed government and public alike, and prompted calls for increased regulation of the industry.

The oil industry is facing a new era in its relations with the government. Perhaps the primary problem of the next few years will be to evolve a relationship that will allow the industry to find and produce new supplies at prices consumers can afford to pay, and one that will allow individual companies to function as profitable private competitive concerns.

INDUSTRY BACKGROUND

The oil industry is usually broken down into four segments: exploration and production; refining; transportation, and distribution or marketing.

The companies in the field are usually classified as either "majors" or "independents," though there are no precise definitions of these categories.

Major companies are often viewed as those that are "integrated," that is, involved in all four aspects of the industry—production, refining, transportation and marketing. Independents are viewed as those operating solely or primarily in one area.

In terms of sales, the 10 largest oil companies are:

Company	1973 Sales[1] (in thousands)
1. Exxon	$25,724,319
2. Texaco	11,406,876
3. Mobil	11,390,113
4. Gulf	8,417,000
5. Standard of California	7,761,835
6. Standard of Indiana	5,415,976
7. Shell	4,883,808
8. Continental	4,224,004
9. Atlantic Richfield	3,982,585
10. Tenneco	3,910,472

Oil represents a major segment of American industry. In 1972, 23 companies had revenues of $250 million or more.[2] In 1971, the top five companies earned profits of more than $5 billion. The next 25 earned more than $2 billion.[3] Net income on sales was 7.2 percent.[4-6]

Total U.S. crude oil consumption in 1973 was 17.3 million barrels per day, of which approximately one third was dependent on Foreign sources. Estimates for 1974 indicated that while the total daily consumption was down slightly, the level of imports remained steady.

The oil industry has had trouble keeping pace. U.S. energy consumption per capita increased at 2 percent annual rate between 1950 and 1970. But between 1965 and 1970, the rate was 3.7 percent, and in the period 1971-72 it hit four percent.[7]

The industry did not recognize the trend immediately, and was not able to expand its capacity rapidly enough when it did. Shortages of fuel oil began to appear, and spot gasoline shortages occurred in the summer of 1973.

These shortages, primarily the result of inadequate refining capacity, were not serious, but they gave the public its first warning that energy might be running short.

Then came the October 1973 Middle East war and the Arab embargo, which cut off nearly a third of the United States' oil imports. Shortages, particularly of gasoline, became acute, and the public began increasingly to look to government for a solution.

The oil industry has historically operated as a private enterprise within the framework of laws and regulations imposed on all American business.

But unlike many other businesses, oil companies produce products that are essential for both the domestic economy and the national defense. Thus the industry has come under closer scrutiny and has been subjected to requirements designed to ensure adequate supplies should war occur.

One such set of requirements was the import quotas imposed during the late 1950s. Designed to protect domestic operations from foreign oil, which could be produced very cheaply, the quotas sustained a high price, and domestic exploration and development activity remained high.

During the late 1960s, pressure from East Coast consumers led to exempting of residual fuel imported into that area. This, plus numerous special exemptions, made price increases on domestic production impractical and exploration declined.

Finally, in early 1973, growing shortages led to the abolition of the quotas.

The effect of the quotas in encouraging, or failing to encourage, domestic production is now the subject of debate. However, as one analyst[8] puts it: "The domestic political unwillingness to support what then seemed to be exceptionally high prices for the domestic energy industries has certainly had an impact on domestic supply."

Thus, government regulation served to give the consumer the worst of both worlds: elevated prices in a climate of uncertainty sufficient to curtail investment in domestic exploration.

It is against this background of growing demand, uncertainty of supply, and the as-yet-undetermined national energy policies and government role in assuring supply that the following issues are identified and conclusions and recommendations presented.

OIL INDUSTRY ISSUES

Major issues facing the oil industry fall into five general categories: decline of currently available supplies; expansion of facilities; alternative sources; tax, price and other government fiscal regulation; and competition.

Decline of Available Supply

Domestic. For the past 15 years, crude oil discoveries have lagged progressively further behind consumption. Not since 1959 have discoveries in a single year been adequate to replace that year's consumption. Discovery of new reserves has declined by 50 percent in the 1960s. From 1961 to 1965, new reserves equaled 65 percent of consumption; from 1969 to 1973, the figure declined to 38 percent.[9]

Only improvements in methods of recovery from known reserves prevented further depletion of domestic supplies.

Domestic exploration has increased sharply in recent months, as Congress has allowed the price of newly discovered oil to reflect increases in the world price. However, government-imposed price restraints have continued to keep the U.S. price below the world market's.

Foreign. Through the 1950s and 1960s, foreign oil, particularly from the Middle East, was vastly cheaper to produce than domestic, and was a continuing source of downward pressure on domestic prices.

By 1973, however, three things had happened.

1. Domestic consumption had hopelessly outstripped domestic production, so that the leverage of the type used against Iran's attempt in the early 1950s to raise prices was gone.
2. Political tensions had reached levels that forced even the most conservative Middle East governments to join in the use of the "oil weapon" against the United States.
3. Expansion of production capacity, coupled with higher prices, pushed revenues to the limits of most producer countries' abilities to absorb them. When that point is reached, the producer country loses incentive to expand capacity. In other words, "A stable flow of oil revenues [is] no longer critical to economic growth in the short run. OPEC countries are comfortable in the knowledge that surplus monetary reserves could see them through any production cuts."[10]

Thus the OPEC (Organization of Petroleum Exporting Countries) nations had the leverage and the incentive to raise prices and impose the embargo, and they could do so without damage to their own economies.

It is this last factor that may prove most significant in the long-run foreign oil prospects. Although some exporting countries have ample capacity to absorb additional revenues, none of those with large reserves do. Therefore while such countries as Algeria, Indonesia, Venezuela, Iraq and Iran[11] would welcome additional cash, their reserves are not sufficient to supply the oil needed.

On the other hand, Saudi Arabia and Abu Dhabi, for example, have extensive reserves but apparently find revenues already adequate.

Estimates of U.S. oil import requirements range from about five million barrels a day in 1985 to about 20 million barrels, depending on such factors as conservation and development of alternative fuels. Whether or not imports will be available in the amounts called for at the high end of the range is not certain. Certainly the Mid-east countries cannot be relied upon for the entire amount, and the potential of other countries is not now clear.

In addition, the United States currently has not a single port capable of handling the supertankers that have become the major mode of transporting foreign crude oil. Today oil is brought to the Carribbean, unloaded, refined and/or loaded on smaller ships and brought into the country. This obvious inefficiency could be eliminated by construction of supertanker ports. A number of proposals have been advanced, including submerged pipelines that would allow the big ships to unload into buoy-like terminals several miles out to sea, but so far there has been little action. Environmentalists fear spills and other environmental damage, while studies show that modern, well-run tanker ports are very clean and cause little damage.

Expansion of Facilities

The public's first warning[a] of an impending energy crisis was the shortage of fuel oil that appeared in the winter of 1971-72. Similar shortages also occurred the next winter and were followed by gasoline shortages in the summer.

The problem stemmed from a refining capacity shortage, which in turn was exacerbated by the government's oil import program. Existing refinery capacity also could not be used correctly because of mis-allocations forced by price control decisions. The import program created uncertainty about future crude oil availability in this country, and then when import quotas on refined products began to be lifted in the mid-1960s, companies found it more economical to build refineries abroad.

With the abolition of the import program, the industry has sought to expand its capacity.

Expansion attempts have been accompanied, however, by a "not in my backyard" public reaction, particularly along the East Coast.[b] Heavy

---

[a]To which it responded! Oil companies had long been warning the public of the impending problem. The fuel oil shortage was the first disruption which "caught the public's ear."

[b]The following is a list, compiled by the Treasury Department, of major refineries that have been prevented or held up by environmental considerations:

Shell Oil Company	Delaware Bay	150,000 barrels per day
Northeast Petroleum	Tiverton, R.I.	65,000 barrels per day
Steuart Petroleum Company	Piney Point, Md.	100,000 barrels per day
Hampton Road Energy Co.	Norfolk, Va.	180,000 barrels per day
Atlantic Richfield	Machiasport, Me.	200,000 barrels per day
Metropolitan Petroleum Company (Pittston)	Eastport, Me.	250,000 barrels per day
Fuels Desulfurization, Inc.	Riverhead, L.I.	200,000 barrels per day
Main Clean Fuels	Searsport, Me.	200,000 barrels per day
(Main Clean Fuels is a subsidiary of Fuels Desulfurization, Inc. and the refinery proposal at Searsport is the same proposal which was turned down at Riverhead Long Island)		
Supermarine, Inc.	Hoboken, N.J.	100,000 barrels per day

restrictions, even outright bans, have appeared, impeding rapid development of needed capacity. Such restrictions have made planning difficult, particularly where they have been imposed after a company has invested money in site-acquisition and construction.

The experience of the Shell Oil Company in Delaware is illustrative.

With the easing of oil import restrictions, Shell, the only major oil company without a refinery on the East Coast, found its competitive position worsening in that area because of the cost of transporting its products there.

In mid-1969, Shell began to intensify plans for a refinery to be located on 5,000 acres the company had acquired in Delaware in 1961. In January, 1970, however, Governor Russell Peterson declared a moratorium on coastal area development, including the property Shell held as its proposed refinery site. At the same time, the governor appointed a Task Force on Marine and Coastal Affairs to make a study and formulate recommendations for future use of the state's coastal lands.

In December, 1970, Shell testified before the task force concerning its refinery plans. This was the first public exposure of the plans, and focused widespread attention on them.

Shell found that the public's image of a refinery was based on older facilities in the Delaware River Basin that were both unsightly and sources of pollution.

Even though it planned to use the latest technology to meet all physical and environmental standards–and had even acquired additional land for use as a buffer around the plant–the company was unable to erase the unfavorable image.

In June, 1971, based partly on the task force's recommendations, the state legislature approved the Coastal Zone Act prohibiting heavy industrial development, including oil refineries, within specified areas. Shell's site was in such an area.

At the time of Shell's planning, the energy crisis had begun to emerge, but had not yet attained public credibility. Now, following the shortages of 1973-74, Delaware's Coastal Zone Act has come under attack in the state and may possibly be repealed.

Shell still owns its Delaware site, and company officials say they feel there is some chance they may eventually be able to use it. In the meantime, however, even though the land itself has appreciated in value, the company has lost its entire investment in engineering and other feasibility studies to determine the site's suitability for a refinery. It has also lost the value of the time and effort it put forth in drawing its plans and defending them.

In addition, Shell incurred large intangible costs in that if it had known its Delaware plan would be disallowed, it could have obtained a site elsewhere and proceeded with the refinery. Currently, in fact, the company is obtaining the necessary approvals for erecting a refinery in New Jersey.

Alternative Sources

"The current higher oil prices," according to one student of the world situation,[12] "are improving the economics for many forms of energy so rapidly that soon the primary limits on new energy projects should be the ability to obtain mineral lands and the required various federal and state approvals."

Additional supplies of petroleum can be made available in two ways: expansion of conventional types of reserves, and development of new extraction and synthesization processes.

Conventional Reserves. Conventional crude oil reserves occur in pools trapped within the earth's crust. Today, although most of the easily accessible such pools have been discovered and exploited, they are by no means exhausted.

The North Slope of Alaska and the outer continental shelf are known to contain extensive oil reserves. Technology is currently available to exploit these areas, and the industry stands ready to proceed.

Estimates of recoverable reserves in Alaska range from 19 billion barrels to 70 billion barrels, plus another 30 billion to 60 billion barrels in Alaskan offshore areas.[13]

Another 50 billion to 90 billion barrels are thought to lie in the continental shelf and the lower 48 states.[14]

Thus development of these reserves would provide the nation with major new supplies of crude oil.

Nonconventional Reserves. Crude oil occurs in many places in nonliquid form—notably in oil shales and tar sands. Both of these types of occurrences are plentiful in the Western Hemisphere, and oil shales occur extensively within the United States.

The Green River formation alone—which underlies parts of Utah, Wyoming and Colorado—is estimated to contain 600 billion barrels in formations ten or more feet thick with concentrations of 25 gallons or more per ton of rock. In addition, the same area contains an estimated 1,200 billion barrels in ten-foot formations at concentrations of 15 to 20 gallons per ton. Other parts of the country possess even larger amounts but in lesser concentrations that must await improvements in recovery technology and higher prices before they become economically feasible to develop.

Synthetics. In addition, petroleum products can be created from other materials, notably coal.

The vast size of America's coal reserves is sufficient to provide raw material for almost unlimited quantities of synthetic crude oil. The main obstacles today are economic—conversion plants cost hundreds of millions of dollars—

and regulatory. Much U.S. coal lies on government lands, and in areas where water, needed in conversion, is scarce. In addition, coal production elsewhere is hampered by environmental and safety restrictions. These uncertainties are not conducive to the large investments required.

The issues surrounding these alternative sources are the interrelated ones of economic cost and environmental cost.

The rising price of crude oil has indeed made many of these sources attractive, but the oil industry, confronted with requirements for extensive outlays of capital for research, land/lease acquisition and construction of facilities, needs to know the constraints under which it will be operating.

Recent government auctions of leases for oil shale lands in the west brought bids ranging into hundreds of millions of dollars, demonstrating clearly the willingness of the industry to stake large sums on such new sources.

But industry also is obligated to protect its financial viability. It cannot risk large amounts of capital where it is likely to face long delays and/or escalated environmental requirements. An experience such as Shell's with its Delaware refinery, if repeated on the scale of the investments necessary to develop such things as oil shale and synthetic crude oil, would damage even large companies severely.

Corollary to the question of developing new energy sources is the question of shared technology among oil companies. Close cooperation among industrial giants has become suspect to the American public, and joint efforts and pooling of resources now run the risk of antitrust action. At a time when technological advances are more desirable than ever, and the associated costs higher, many in the oil industry feel it is time for the government to develop a policy that would free companies to make maximum possible use of their abilities.

Tax, Price and Other Government Fiscal Regulation

Taxes. For many years the federal government has recognized the unique nature of the natural resource industry by granting it a depletion allowance on taxes. The allowance was designed to encourage investment in natural resource industries, and to offset the fact that such resources, by their nature are non-renewable and inevitably exhausted.

U.S. tax policy has also recognized that the oil industry, in developing resources and producing crude oil in foreign countries, is subject to large taxes abroad. The American government has, therefore, allowed companies to credit taxes paid to other governments against taxes due to the United States.[c]

These provisions are objects of controversy. Critics feel they unduly reduce oil company taxes and contribute to inflated profits. The same tax credit is available to all U.S. companies doing business abroad. Of special

[c]These credits cannot offset U.S. taxes on income earned in the U.S.

benefit to the oil industry is a special Treasury Department ruling that allows royalties paid foreign governments to be treated as taxes. Defenders believe these provisions are necessary to attract capital to enterprises with a higher than average risk factor, and that eliminating them would further impair the ability of the industry to develop new resources and facilities.

Prices. The petrochemical and refining sectors of the oil industry remain under government price controls. The controls, which continue even after expiration of other wage-price controls, caused considerable dislocation in the industry in 1972 and 1973, particularly in refining.

Competition

Integration. The major American oil companies are "integrated," that is, they explore for and produce crude oil, refine it, and market the products. There are 18 such companies.[15] Out of the 1973-74 energy crisis have come calls to dismantle the integrated companies. The charge is that they stifle competition and manipulate supplies to drive up prices. According to a Federal Trade Commission staff study, "The major firms seek to consolidate market power by various exclusionary tactics. These firms basically attempt to sharply limit the supply of crude oil available to independent refiners and refined product available to independent wholesalers and retailers . . . The resulting system endangers existing independents, makes new entry [into the industry] difficult or impossible, and yields serious economic losses to American consumers."[d]

The industry replies that there is vigorous competition both among majors and between majors and independents and that integration supplies both the scale necessary for economical operation and the concentration of capital necessary to plan and build the $100 million to $500 million facilities necessary in the industry today.

And, according to Stephen A. Wakefield, assistant secretary of the interior for energy and minerals:

> In the new environment of high prices and scarce supplies . . ., the prospect is for a relentless competition for every available barrel of oil by consuming nations all over the world. And in this competition, the United States will need every tool and lever it can apply to promote the flow of oil in our direction. Our principal asset in this regard is the in-place network of transportation, refining and distribution facilities of the integrated oil companies. . . . And dismantling of this domestic structure on our part would so

[d] The Treasury Department staff analysis of the preliminary Federal Trade Commission staff report on its investigation of the petroleum industry, SN93-18 (92-53) is helpful as a balance to the FTC report.

> weaken the attractiveness of U.S. markets that this act in itself would assure that the United States stood at the end of the line of claimants in a world where energy scarcity is likely to be the normal condition for many years to come.[16]

Diversification. Oil companies today have major holdings in other energy fields. The largest coal company is owned by an oil firm. Many also are involved in the production of nuclear fuels, and 70 percent of U.S. natural gas is produced by oil companies.

Here again critics suspect manipulation and restraint of trade. Industry officials cite their companies' technical expertise as well as the simple rational business practice of trying to assure future supplies –e.g. acquiring coal as a potential source of synthetic crude oil. According to Charles E. Spahr, chairman of Standard Oil of Ohio:

> . . . The traditional coal industry was grossly undercapitalized. The return on borrowed capital was, and had been for many years, far below that for total industry in the United States. It would be required . . . . Sohio would be able to provide substantial capital . . . .[17]

Distribution and Marketing. During the Arab embargo, a number of independent gasoline marketing concerns, which had been buying excess supplies from larger companies, found themselves unable to obtain the amounts they had expected as desired. Many perceived the shortage as a conspiracy to drive them out of business.

In fact, gasoline marketing is a large and diversified segment of the American economy. There are more than 13,000 independent wholesalers in the industry and more than 200,000 retailers,[18] most of whom are independent businessmen.

The Arab embargo was a difficult period for the industry, and many marginal businesses did indeed find themselves unable to survive. But this was not confined to independents. Major concerns were also forced to close outlets and cut back retail operations.

Today independent suppliers and retailers are competing vigorously with major companies for reserves and markets. Some 75 companies bid during two recent offshore lease sales, and almost all retailers, regardless of affiliation, are able to obtain and sell sufficient quantities of gasoline to assure their continued viability.

CONCLUSIONS

1. Energy supplies consistent with continued U.S. economic growth cannot be provided by domestic production at current rates.
2. Internal constraints will probably limit many countries' willingness to develop their oil resources as U.S. demand mounts.
3. Additional supplies are available from other U.S. and Western Hemisphere sources, but large investments and environmental risks are involved.
4. Supplies from nonconventional sources are also available, but development of these will require firming up the long-term profit and price outlook, justifying large investments and some environmental risks as well.
5. Synthetic sources are of enormous potential because of the vast supplies of coal possessed by the United States.
6. Government economic regulation of the industry has produced mixed results. Tax incentives have spurred development, buy price controls have led to dislocations and will inhibit development of additional supply for the United States.
7. Environmental requirements have sometimes imposed undue hardships on industry, as noted in the Shell-Delaware case, and, as noted in the same case, have not always been based on adequate understanding of what is really being proposed.
8. Concern over antitrust inhibits joint planning actions and research and development programs that may be in industry's and the public's interest.
9. Excessive dependence on imports, in the current international situation, makes pricing dependent on foreign initiatives; world and U.S. prices are not within U.S. control. This impedes planning of supply from alternate sources such as tar sands, shale and coal.

RECOMMENDATIONS

1. Every effort should be made to encourage domestic exploration and development. Government controls that inhibit development of additional supplies should be reviewed in light of this need.
2. Development of new sources should be encouraged. Access to government lands should be granted, and issuance of permits for construction of processing plants, pipelines and other necessities should be facilitated.
3. Research should be encouraged, so that new sources may be found and costs of using known ones reduced.
4. Oil company revenues must be allowed to reflect additional costs of developing new sources. This may be done through competitive pricing in the free market or through government subsidies or a combination of both.

5. The nation's timetable for achieving environmental goals should be revised to reflect the need for meeting energy needs. The oil industry supports environmental goals, but both environmental goals and energy needs must be viewed as different aspects of the same problem, and trade-offs between them allowed.

NOTES TO CHAPTER TWO

[1] Fortune 500.

[2] Treasury department staff analysis of preliminary Federal Trade Commission staff report on its investigation of the petroleum industry, SN 93-18 (92-53)..

[3] *The American Oil Industry. A Failure of Anti-Trust Policy,* by Stanley H. Ruttenberg and Associates for the Marine Engineers' Beneficial Association, 1973.

[4] Ibid.

[5] The Arab Oil Embargo and Its Impact on Winter Fuel Shortages, Congressional Research Service, November 1973, pp. 3-19.

[6] *OCS Oil and Gas–An Environmental Assessment,* Council on Environmental Quality, April 1974.

[7] Ibid.

[8] *The World Crude Oil Outlook,* William D. Witter, Inc., March 1974.

[9] Ibid.

[10] Ibid.

[11] Ibid.

[12] Ibid., p. 27.

[13] *OCS Oil and Gas–An Environmental Assessment, Council on* Environmental Quality, April 1974.

[14] Ibid.

[15] Federal Trade Commission, Preliminary Staff Report on Its Investigation of the Petroleum Industry, pursuant to S. Res. 45, #93-15, p. 12.

[16] Testimony before Senate Committee on Interior and Insular Affairs, November 28, 1973.

[17] Ibid.

[18] Binsted, Charles, President, National Congress of Petroleum Retailers, testimony before Senate Committee on Interior and Insular Affairs, December 5, 1973.

Chapter Three

Electric Utilities

In 1973, 26.2 percent of the primary energy supply of the nation (oil, gas, coal and nuclear) was converted to and used in the form of electricity. It is estimated that 50 percent of the primary energy supply will be converted and used as electricity in the year 2000.[1]

Much of the increase can come from the substitution of electricity for gas and oil, as those fuels become more expensive and scarce; and will be provided by increased use of coal and nuclear power.

Thus the electric utility industry, which is just completing a decade of unprecedented growth, faces another period of immense expansion. Problems that have been troublesome in recent years must be resolved lest they become major impediments to an adequate energy supply. These problems include the raising of capital, obtaining fuels, and meeting service and environmental requirements.

INDUSTRY BACKGROUND

Commercial electric power has been sold in the United States since the late 1870s. From tiny beginnings such as Thomas Edison's Pearl Street Station which served 59 customers in a one-mile radius in New York City, the industry has grown into a major segment of the American economy.

Today more than 3,100 utilities provide some 80 million customers with more than 1.7 trillion kilowatt hours annually, and the numbers are increasing steadily.[2]

In 1973, the industry had a total installed generating capacity of 438,493 megawatts, up from 399,606 in 1972. Its total sales to customers in 1973 were 1,703,203 million kilowatt hours, up from 1,577,714 million the year before. The industry served 67,314,000 residential customers in 1972 and

69,438,429 in 1973; 8,200,033 small light and power customers in 1972 and 8,361,847 in 1973.

To provide this service the industry had in service more than 475,589 miles of over head electric lines (22,000 volts and above). It used 386 million tons of coal in the 12 months ending November of 1973, 565 million barrels of oil, and 3,754 billion cubic feet of natural gas. Total fuel consumption, including nuclear power, was the equivalent of 723 million tons of coal.

Investor-owned utilities had, at the end of 1973, commitments of $55.9 billion for capital outlays. These utilities totaled $27,400 million in revenues in 1973 for a net profit of about $5.02 billion.[3]

The electric utility industry is divided into two basic categories: investor-owned companies and publicly controlled utilities. The investor-owned companies, though fewer in number, produce slightly less than 80 percent of the nation's electricity. These companies, numbering about 275, are generally very large, fully integrated companies, that is, they generate, transmit and distribute power. Together they serve 61 million customers.

Publicly owned utilities number 2,920, serving 17 million customers. Generally smaller–though the Los Angeles Department of Water and Power serves 1.1 million customers–three quarters of them are in communities of 10,000 or less. Many are involved only in transmission and/or distribution.[4]

Regulation is almost as old as the industry, dating back to 1907 when New York and Wisconsin set up the first state utility commissions to regulate rates. Today such regulation is nationwide.

Until recently, however, the trend in electric rates had been down, so that regulation often involved little more than an agreement between utility and regulator to pass on to the consumer new economies brought about by major technological advances.

In the 1960s, this changed. Sharply increased consumption coupled with rising costs–both of operation and of capital improvement–brought about a need for higher rates. Environmental restrictions on fuels, on generator sites, on power line locations, have all increased costs drastically.

Many utilities, particularly in the urban areas of the East Coast, have been forced by environmental concerns to use low-sulphur imported fuel oil. Actions of the oil-producing nations in raising prices have put a severe strain on many companies; this, plus sharply higher rates brought on by fuel-cost pass-throughs have thrown the problems of the utility industry into sharp focus.

UTILITY INDUSTRY ISSUES

Issues facing the utilities industry today may be loosely grouped into four major categories: capital and revenue requirements, availability and use of fuels, environmental and safety regulations–including construction permits, and pooling and mergers.

Capital Revenue Requirements

For the vast sums required to construct the facilities necessary to keep pace with rising demand, investor-owned and many government-owned utilities have traditionally turned to the public for funds, either through bond issues or stock.

Today, faced with rising operating and fuel costs, very high new facility costs and, in some cases a new uncertainty about sales, many utilities find their financial situations difficult. At least one, Consolidated Edison of New York, is in a crisis.

Many factors bear on utilities' financial situations, but several are particularly important

"Coverage." In seeking to borrow money, companies generally are not permitted to sell additional interest-bearing securities unless their annual income before interest charges and all income taxes is about twice[a] annual interest payments. Thus, the companies' ability to raise new capital is contingent upon their ability to generate sufficient revenues and profit to meet the coverage requirement.

Until recently utility revenues were generally adequate to meet the coverage requirement easily. Today, however, this is no longer the case, and while few companies have actually reached the point where they may not legally offer new securities, investors have become more aware of the companies' deteriorating situation and bond ratings have plummeted. In an already tight interest market, some companies are being forced to postpone necessary borrowings and the installation of new facilities.

Price/Service Regulation. The companies' problems arise in large degree from their unique position as a regulated industry. The regulated electrical utility has a legal obligation to supply all electric power required by its customers in its service area, and it has a legal obligation to build the plant required to provide that service.

Where a nonregulated industry finds it uneconomical to sell a particular product or to invest in a proposed expansion of plant, it has the option of cutting its losses and discontinuing unprofitable sales or foregoing unprofitable expansion. The regulated electric utility cannot do so. It must supply its product even where it believes the price allowed makes the sale of new services or construction of new plant uneconomical.

Also unlike a nonregulated industry, the regulated electric utility cannot increase its prices immediately to recover costs as they are incurred. Even where the regulatory agency agrees that the increase is warranted—because of the delay that inevitably surrounds agency decisions and because such decisions are generally based on prior year costs—the utility, in a period of inflation, will always be a year or more behind in catching up with cost increases.

---

[a]Varies in range of 1.75 to 2.50, based on specific indenture terms.

Further, an electric utility has little or no control over a major part of its total costs. These include the cost of capital and of taxes, which run 50 percent or more of total costs. Also, increasingly rigid and constantly changing environmental requirements add even more to capital and operating costs.

Regulation was first imposed for the health of the industry. Because it was felt that normal competition was not appropriate in a capital intensive industry like electric power, companies were franchised by the government and regulatory agencies set up to ensure that the businesses obtained a reasonable rate of return and the public was charged reasonable rates.

To paraphrase Herbert Cohn, vice chairman of American Electric Power Co: The principal function of regulation should be to assure, to the fullest extent possible, that the regulated utility provides reliable and adequate service to the public. Where the regulator is convinced that increased rates are essential to enable the utility to finance the facilities required for this purpose, such an increase should be authorized promptly. The objective of reliable and adequate service is an essential element in furthering the public interest, and regulation at all levels–administrative, legislative and judicial–should help to further this objective–or at least not frustrate it.

Cohn also urges that the environment be taken into consideration, but that it be weighed along with other aspects of the problem. "What we do urge is that these problems be approached with a full consideration of *all* aspects of the public interest, and with a greater sense of perspective, to reach a rational and balanced judgment–the governing consideration to be what will best further the overall public interest."

Today regulatory agencies, far from ensuring the health of the electric utility industry, are for the most part bleeding it to death. The case of Consolidated Edison of New York (Appendix H) provides a concrete illustration of the following problems:

Regulatory Lag. Rate adjustment procedures are very long, and are usually based on a "test year," which the agency examines as a basis for deciding whether rate increases are justified. Unfortunately, in a period of high inflation and rapidly rising interest rates, the test year is no longer relevant by the time the new rates go into effect.

Deficient Rates. In addition to the delays involved, regulatory agencies do not always grant increases that would be adequate even if implemented right away. Because of political pressure, poor judgment or simple miscalculation, regulatory agencies often grant rates that do not provide a rate of return adequate to make the utility attractive to the financial community. This has adverse effects on both the company's stock and its bond offerings–the most basic ways by which it raises capital for expansion and improvement.

Future Repercussions. Inability to raise large amounts of new capital, particularly in a period of growing consumption and/or rapid technological advance, has serious implications for a utility's ability to provide adequate and reliable service in the future. Apparent savings to the consumer today may mean larger costs to him in the future either as his service becomes unreliable or his rates are at last forced much higher as the utility plays technological "catch up."

Availability and Use of Fuels

The electric utility industry has traditionally been a hard and effective bargainer in the fuel market, with a history of keeping costs well in line and profits up.

Today, however, two factors have intervened in this process, both tending to put the industry at a disadvantage and driving costs up sharply.

First is the imposition of environmental requirements that eliminated, or threatened to eliminate, traditional supplies of high sulfur coal and dictated use of more expensive and/or harder-to-get low sulfur coal, oil and natural gas. Uranium remains economical in itself, but is a choice only when new facilities are considered, and accompanying any nuclear projects are problems of site selection, licensing, safety and environmental regulations, the administration of which is costly and time-consuming.

Second is the sharp increase in the price of fuel oil, especially imports upon which many East Coast companies had become dependent.

The two factors compound one another. Environmental requirements force the utilities to buy oil despite the high price, thus preventing the utilities from boycotting or switching to coal.

Legislation, as mentioned earlier, is being considered that would allow utilities to burn more coal, but at the same time existing clean air laws impose even stiffer restrictions over the next two years that will tighten the fuel market even further.

Natural gas is nearly ideal as a fuel—cheap, clean and easy to handle—but supply is already inadequate to meet demand and at present rates of development this situation is not likely to change.

The United States does possess large quantities of low-sulfur coal in the western part of the country, but transporting it to the major consuming regions is expensive, and conversion to electricity or other fuels on site faces environmental and other problems (see *Coal*).

Questions concerning availability of fuels are important to utilities. Construction of new facilities involves vast sums, as does purchase of fuel over the life of the facilities. Thus a company needs to know as far in advance as possible what will be available and under what conditions it can be used, so that it can plan a plant to make use of the most economical fuel and so that it does not find itself faced with having to back-fit pollution control equipment it had not expected to need.

Environmental Regulations

In general, environmental regulations are regarded by the electric utility industry as part of an emerging national environmental ethic to which the industry can adapt with enthusiasm. Problems arise, however, when regulations are imposed without thorough evaluation as to feasibility and, pragmatically, to costs versus benefits. Particularly, the industry feels that a majority of the public has not been made aware of what the costs are likely to be.

Meeting environmental regulations adds significantly to utility costs. Regulations influence both the location of proposed facilities and, in the case of generators, the kind of fuel used. One particular regulation, the federal requirement for an environmental impact statement, has given opponents a mechanism to delay, and sometimes defeat, new facilities.

A decision by the U.S. Circuit Court of Appeals[5] that the National Environmental Policy Act of 1969 requires an environmental impact statement by the Atomic Energy Commission before licensing a nuclear power plant would, the Federal Power Commission estimates, add an entire year to the lead time for nuclear generating plants.

Subsequent decisions have widened the scope of NEPA to include the disposal of wastes—wastes defined to include heat from steam-electric power plants—so that since 1971 when the first decision, in the Calvert Cliffs, Md., case, was handed down, a whole new spectrum of requirements has been imposed on the industry.

Restrictions have also been applied retroactively, which also imposes heavy burdens on industry. Cooling towers, for instance, fitted on generators not originally designed for them are very costly. Gordon Corey of Chicago's Commonwealth Edison Company notes that one of his company's plants would lose 8 percent of its capacity and as much as $90 million would be added to construction and operating costs if it were retrofitted with cooling towers.

In addition there are those who believe that delay in and of itself will slow energy growth and thus benefit the environment. According to Irving Like, a New York lawyer active in opposing at least one major power plant in the east, " . . . A losing environmental cause is worth fighting for because it adds to the ecological enlightenment of the public. It is even possible that a prospective polluter may be induced to abandon its plans or at least improve upon them if it knows that its project will provoke an environmental challenge."

Pooling and Mergers

Of the more than 3,100 electric utility companies in the United States, many are small companies involved in only one or two aspects of the business. Some are distributors only; some generate and transmit only.

There is a wide range of sophistication, and relationships among them are diverse.

The relationships, especially among investor-owned companies, are further complicated by what is loosely referred to as "Pooling." This term covers a variety of arrangements between systems to obtain various cost-reducing and service-reliability advantages of interconnection. The term power pool refers to the interconnection of two or more systems for two-way exchange of power.

The small companies have traditionally argued that their continued existence demonstrates their economic viability. But as fuel supplies become more expensive and less reliable, and new equipment more expensive and more sophisticated, many in the industry feel that economies offered by large holding-company type organizations are necessary.

Larger organizations offer economies of scale, they believe, and can more readily bring together and use the managerial, financial and technological resources required to master the larger and more complex problems the industry now faces. Decision-making is simplified when only one entity is involved. Pools, on the other hand, involve compromise among the participants, and these compromises may lead to inefficiencies in planning and providing new facilities.

Small companies, out of self-interest, are likely to resist, and antitrust questions are involved. But in the present situation, many observers feel mergers should be encouraged.

The Public Utility Holding Company Act of 1935, under Section 30, directs the Securities and Exchange Commission to make studies of the industry for the purpose of recommending the "type and size of geographically and economically integrated public utility systems which, having regard for the nature and character of the locality served, can best promote and harmonize the interests of the public, the investor and the consumer." Further, the act specifically recognizes that the appropriate size of integrated systems is to be determined in light of the current state of the art and contemplates increasingly larger systems as warranted by advancing technology.

CONCLUSIONS

1. Electric utilities are a capital-intensive industry requiring large amounts of money to build and maintain high-quality equipment necessary to supply growing demands.
2. Privately-owned companies must maintain a sufficient return on investment so that they may legally raise more capital and so that their offerings will attract investors.
3. Capital, fuel, operating and other costs have been rising much more rapidly in the past three years than ever before.

4. Regulatory agencies are under pressure to resist rate increases; they must also meet numerous requirements for hearings, reviews and other procedures before granting rate increases. Clearly, necessary rate increases are often outrun by events.
5. Environmental requirements on the one hand and supply restrictions on the other are eliminating traditional utility fuels while making others far more expensive.
6. The nation's approach to achieving its environmental goals remains in flux, making planning difficult for utilities and sometimes forcing expensive last-minute alterations in new equipment.
7. Environmental statutes have provided interventionists with a method of delaying proposed facilities for periods that can run into years. Such delays markedly increase costs.
8. Development of commercial atomic power has not progressed as rapidly as predicted; it is now beginning to make significant contributions to generating capacity, but major problems remain.
9. Environmental and health and safety regulatory procedures seriously delay installation of needed generating facilities, especially nuclear ones.
10. Siting of electric generating stations and bringing them into production is increasingly difficult even as the need for more electricity to offset natural gas and oil shortages becomes more pressing.

RECOMMENDATIONS

1. Rate-setting processes should be streamlined to respond more rapidly to industry needs for rate adjustments.
2. Rates must be allowed to rise to keep pace with costs such that utility companies can earn the rate of return on investment allowed by law. Failure to permit this will lead inevitably to deterioration of service and could lead to the collapse of utility companies.
3. Environmental goals should be balanced against energy costs and demands; fuel restrictions and equipment requirements mean higher costs, and an informed public must decide how much it wishes to pay.
4. Clear-cut environmental and health and safety standards should be set, so that a company can take them into account in its planning. Actions by the appropriate regulatory agencies should assure that a project go forward on an orderly basis.
5. Research, particularly toward full utilization of atomic energy and pollutant-free fossil fuels, should be stepped up. Joint research and development, as through the Electric Power Research Institute, should be continued.
6. Utility rates should reflect the cost of providing service to the different customer categories.
7. The studies of the industry called for under Section 30 of the Public Utility Holding Company Act should be activated.

NOTES TO CHAPTER THREE

[1] Edison Electric Institute.

[2] Green, Mark J., *Deciding on Utilities: Public or Private?* New York Times, May 26, 1974.

[3] Data compiled for the author by the Edison Electric Institute.

[4] Green, op. cit.

[5] Calvert Cliffs Coordinating Committee, et al. v. U.S. Atomic Energy Commission, et al. U.S. Circuit Court of Appeals for the District of Columbia, June 23, 1971.

Chapter Four

Gas

Natural gas has been, until recently, the fuel of choice for home and industry in many areas of the United States. Clean and efficient to burn, available by pipeline at regulated low prices, gas rapidly increased its share of the U.S. energy market in the post-World War II years, climbing from 13.7 percent in 1947 to 32.3 percent in 1972.[1]

The principal problem with gas is that there is not enough of it. Originally recovered largely as a byproduct of oil production—and even burned off as waste—gas is now an exploration and production industry of its own, with risks and costs increasing accordingly.

Rates have, however, remained low due principally to federal regulation; in 1972 the average cost to utilities for gas was 31 cents per million Btus, compared to 82 cents per million Btus for distillate oil, 40 to 68 cents for residual oil, and 34 to 38 cents for coal. (Coal and oil prices range depending on sulfur content.)[2] As a result, gas exploration has declined, as has development of new reserves.

In addition, the disparity is increasing as oil prices, particularly for imports, climb and coal producers struggle to raise production.

Today natural gas is in short supply. Distributors in many areas are unable to provide service to new customers. But the Federal Power Commission, the main regulatory agency in the field, has remained reluctant to allow large rate increases even though the producers insist such raises are essential as incentives to locate new supplies.

There is a trickle of alternate supply, in the form of imported liquefied natural gas (LNG), and ambitious projects have been proposed involving massive imports of LNG and gasification of coal. However, these supplies would be very costly compared to present natural gas prices and involve resolution of major environmental and in the instance of LNG international political and monetary problems.

INDUSTRY BACKGROUND

Natural gas consists mainly of the very simple hydrocarbon, methane. It is formed the way crude oil is—from organic material buried under great thicknesses of sedimentary rock--and the two are often found together.

Much natural gas is used directly—that is, burned by the end user for its heat. It also has been widely used by electric utilities, initially because it was cheap, and more recently, where it is still possible for utilities to buy gas for boiler use, it is popular where pollution controls are strict.

Industry other than utilities is the single largest user of gas, accounting for about 40 percent of consumption. Residential users consume about 25 percent, and electric utilities about 20 percent.[3] As a fuel, it is valued for its high heat value, and especially because it is free of the problems associated with many other energy sources: it produces no sulfur oxides, no particulates, no solid waste and no nuclear radiation, and it is obtained from wells so that it is free of the problems of strip or underground mining.

In 1972, the United States produced 22.9 trillion cubic feet of natural gas, out of a total proven reserve of 266.1 trillion cubic feet. In addition, slightly more than 1 trillion cubic feet were imported and 79 billion cubic feet exported (most as LNG from Alaska to Japan).

Total U.S. reserves are estimated variously at between 1,550 trillion and 2,950 trillion cubic feet.[4]

The West South Central states[a] account for approximately 80 percent of U.S. natural gas production, and are also leading consumers, followed by the East North Central area.[b]

The natural gas industry is divided into three segments: production, transmission (usually pipelines) and distribution.

The producers, usually oil companies, recover the gas from wells and sell it to the pipeline companies. The pipelines deliver it "to the city limits," where local gas companies, the distributors, sell it to retail customers.

Gas is delivered through 961,000 miles of pipeline to more than 43 million customers in the 49 continental states. The number of domestic oil and gas producers has been estimated at more than 30,000, with sales of more than 3,700 of them subject to Federal Power Commission regulation. In 1971, 22 producers together sold more than 1000 billion cubic feet of gas—71 percent of the gas shipped through interstate pipelines.[5]

There are 106 regulated interstate pipeline companies, 33 of which are considered major. Distribution companies total more than 1,500.[6]

---

[a] Arkansas, Louisiana, Oklahoma and Texas

[b] Illinois, Indiana, Michigan, Ohio and Wisconsin

Leading Companies in each segment of the industry are:

Largest Natural Gas Producers 1971

Company	*Sales to Interstate Pipelines (billion cubic feet)*	*Market Share (percentages)*
Humble*	1,334.4	10.5
Shell	783.7	6.0
Amoco	780.4	6.0
Gulf	751.0	5.8
Phillips	724.2	5.6
Mobil	683.6	5.3
Texaco	609.8	4.7
Union	549.9	4.3
Atlantic-Richfield	500.1	3.6
Continental Oil	466.2	3.4
Total for Top 10	7,197.	
Total domestic sales into interstate pipelines	13,716.3	
Sales to Canada, Mexico	881.5	

Source: AGA.

*Now Exxon USA.

Largest Gas Transmission Companies

Company	*Sales** (millions)*	*Volume* (million mcf)*
El Paso Natural Gas	$799.2	1,661.4
Columbia Gas Transmission Company	794.8	1,335.7
Tennessee Gas Pipeline Company	593.1	1,293.3
Texas Eastern	492.0	874.2
Natural Gas Pipeline of America	487.1	964.0
Transcontinental Gas Pipeline Corporation	476.9	999.7
Consolidated Gas Supply Corporation	439.7	623.6
Michigan-Wisconsin Pipeline Company	427.3	815.8
Northern Natural Gas Company	419.5	693.6
United Gas Pipeline Company	382.6	820.2

Source: FPC.

*Gas sold for resale, 1973; figures preliminary.

**1972.

Largest Gas Distribution Companies***

Company	Revenue (millions)		Volume (mcf)	
	1972	1973	1972	1973
Northern Illinois Gas Co.	$424.9	$447.6	507,407,800	503,936,800
The East Ohio Gas Co.	366.8	348.3	412,833,307	384,914,918
Columbia Gas of Ohio, Inc.	340.9	335.9	409,600,000	375,700,000
Peoples Gas Light and Coke Co.	310.9	295.3	326,290,621**	276,217,932**
Gas Service Co.	164.2	165.8	293,731,310	287,471,403
Atlanta Gas Light Co.	181.2	201.2	260,885,500**	259,790,400**
National Fuel Gas Co.	236.3	222.8	232,549,000	216,120,000
Columbia Gas of Pennsylvania, Inc.	140.9	142.1	145,121,291	142,507,013
Laclede Gas Co.	135.4	155.6	128,934,099**	139,084,204**
United Gas, Inc.*	97.0	109.2	123,085,746	129,100,947

Source: American Gas Association.

*Now Entex.

**Terms X 100.

***Companies obtaining at least 95 percent of gas operating revenues from sales to ultimate customers and classifying at least 95 percent of mains (other than service pipe) as distribution.

GAS INDUSTRY ISSUES

Additional Supplies

Domestic Exploration. The single largest problem facing the gas industry is the lack of adequate supplies. Domestically consumption has exceeded new discoveries every year but one since 1968. A major reason for this is the rate structure imposed by the Federal Power Commission. As exploration, drilling and other costs have risen sharply, producers have been allowed only small price increases.

Gas exploration is expensive business. According to the Interstate Natural Gas Association of America, the cost of a single rig capable of drilling to a depth of 20,000 feet is more than $1.6 million. Renting such a rig with its backup services runs more than $5,000 a day. And operating an offshore platform can exceed $15,000 a day.

Between 1959 and 1970, according to the INGA, there was a 38 percent overall decline in drilling–from 47,600 wells to 29,460–and a 43 percent drop in exploratory drilling–13,200 wells to 7,700.

Today, drilling and completing an onshore gas well costs, on the average, $170,000; an offshore well in the Gulf of Mexico over $700,000. At the same time, and particularly since 1968, each foot of drilling has turned up fewer cubic feet of new gas. And since 1968, new reserves added have been less than consumption.

In the past two years, the FPC has granted rate increases designed to increase supply. It has granted increases in area rate ceilings ranging from 20 to 40 percent above previous (1969) levels. It has also granted certificates authorizing short-term contracts (up to three years) at prices as high as 45 cents per thousand cubic feet, compared to the maximum area ceiling in Louisiana of 26 cents. Most recently the FPC established a nationwide area rate of 42 cents per thousand cubic feet.

The FPC may also, under its optional pricing rules (section 2.75 of its rules), grant terms for specific contracts if the commission finds them just and reasonable following cost and economic criteria. The commission has used this procedure both to increase area ceilings and to allow some offshore gas to exceed the area ceiling.

In addition, the FPC allowed gas into interstate sales on an emergency basis for 180 days without prior commission approval, which has resulted in some increase in supplies.[c]

Typically, however; FPC moves in this direction have been subjects of litigation, and many have been disallowed.

---

[c]The 180-day emergency sale program was dropped when the areawide rate of 42 ¢ per thousand cubic feet went into effect.

Imports. Other nations possess large reserves of natural gas, but until recently there was no practical way of transporting it to the United States. A modest amount was imported by pipeline from Canada and Mexico, but in more distant lands gas was simply flared off at the oil fields for lack of a way of selling it.

It is possible to liquefy natural gas (at temperatures of –260 degrees F) and transport it in specifically equipped tankers.

This development has opened the way for U.S. use of gas reserves in Central and South America, Nigeria, Indonesia, the Mideast, Algeria, Australia, and the Soviet Union and many American companies are seeking to develop these possibilities.

However, the technical difficulties are considerable and expenses high. Tankers, liquefaction facilities, tanker loading and unloading facilities must be built. In addition, especially where the U.S.S.R. is involved, there are political problems.

Synthetics. Another source of additional supplies is the creation of gas from other hydrocarbon sources, notably coal. The technology is currently available to create gaseous fuels from coal. The economics is uncertain.

The great attraction of synthetic gas is its potentially large supply. If only half the U.S. coal reserves are considered (780 billion tons), at a rate of 16,000 cubic feet per ton of coal, 12,000 trillion cubic feet of gas are potentially available.[7] (Though obviously the total coal converted to gas is likely to be much less.)

Gas can also be synthesized from oil. Technology is currently available for cracking light oils to produce gas, and the process could be used as a source of supplemental supplies. However, installation of conversion plants is costly–plans for about 15-17 plants with a total annual capacity of more than 500 billion cubic feet have been announced at a total cost in excess of $1 billion, according to the American Gas Association–and oil supplies themselves are of uncertain availability.

Other Sources. Recent discoveries of gas in Alaska and promising geological data from offshore indicate that these areas have the potential for large additional supplies of gas.

Alaskan potential reserves are currently estimated at more than 360 trillion cubic feet.[8] However, production costs are high because of the hostile climate, and delivery remains problematical for both technical and environmental reasons. The saga of the trans-Alaska pipeline is well-known, and delays in resolving the dispute have delayed by several years the beginning of Alaskan production.

Currently, several plans have been offered for bringing Alaskan gas to the lower 48 states. One, proposed by a consortium of companies, envisions a pipeline from the gas fields across Alaska and Canada and into the United States.

Another, proposed by El Paso Natural Gas Company, would pipe the gas across to the southern coast of Alaska, where it would be liquefied and transported to the West Coast by tanker.

The outer continental shelf offers even more extensive prospects, but development there faces almost certain battles with environmentalists. A study completed in April, 1974, by the Council on Enviornmental Quality indicated that environmental risks are not excessive for several promising areas, and the gas industry is hopeful that this finding will provide the impetus for a federal policy encouraging expeditious development of the shelf's potential.[9]

The industry feels that many of the fears expressed about possible pollution connected with offshore drilling are overstated. The 1969 spill in the Santa Barbara channel was and is a unique case. There a well was drilled in a highly faulted[d] area, without a full well casing extending below the faulted zone, allowing oil to escape outside the well casing. Nothing like it has happened before or since.

In the Gulf of Mexico, more than 16,000 wells have been drilled over 25 years with no significant environmental damage.[10] The industry feels that a similar record of safety could be expected on the Atlantic and Pacific outer continental shelves, and is anxious to proceed with leasing and exploratory drilling so that this tremendous source of supply can begin to be tapped.

Critics of offshore development point to the possibility of leaks or spills that would pollute water and nearby beaches. While these possibilities exist, it should be noted that natural gas is the cleanest of the fossil fuels, and that increased production will permit reduction in use of other fuels and consequent improvement of air quality.

Price Structure

Underlying all discussions of additional supplies and whether they will be developed, is the basic question of economics—if capital is invested, is there the expectation of a reasonable return.

The price of natural gas for the interstate market is regulated at the wellhead by the Federal Power Commission and to the consumer by state and local public utility commissions. For years the FPC has maintained the price of gas at levels well below the per-Btu prices of other fuels. As costs have risen and prices have not, exploration activity has declined; industry officials have repeatedly told both the FPC and various congressional committees that this trend will inevitably continue unless prices are increased.

In addition, the FPC has sometimes stepped in to reduce prices specified in contracts after the contracts have been given initial approval, thus introducing another element of uncertainty.

---

[d]Fault is a geological term for a crack in the earth's crustal rocks along which slippage has taken place.

The industry favors removal of price regulations so that prices can reflect production costs. This, the industry feels, is the best way to assure adequate investment to develop new supplies.

A third problem with current regulatory procedures is the length of time involved in obtaining decisions.

The FPC also regulates other aspects of the gas industry. Location and construction of pipelines are subject to FPC approval, as are profit levels of the companies. Pipelines are also subject to safety regulations imposed by the National Transportation Safety Board.

Following rate increases granted by the Federal Power Commission in the past 18 months and renewal of proposals to deregulate gas prices, critics made much of the fact that neither the FPC nor other government agencies has any independent data on reserves. Suspicions were voiced that the industry was concealing some of its reserves in order to pressure the government into removing price restrictions. After restrictions were removed, the argument went, the industry would then be able to sell the additional gas for higher profits.

It was suggested that a government agency be formed to investigate independently, so that regulatory agencies would have information from sources other than the industry.

In fact, in the wake of the controversy, an FPC staff study was undertaken and came to the conclusion that industry estimates were, if anything, slightly optimistic.

CONCLUSIONS

1. Gas reserves are declining with respect to consumption at a rate that already has placed some restrictions on gas use, and could, if continued, render supplies inadequate.
2. Vast supplies are believed to exist within U.S. jurisdiction, offshore and on public lands.
3. Large supplies also exist abroad, and technology is now available for transporting these to the United States as liquefied natural gas or other liquid fuel forms such as methanol.
4. Promising processes exist for converting coal and liquid hydrocarbons into gas.
5. Gas has been available at prices below environmentally less desirable fuels.
   a. The price structure has made it more profitable to develop oil than gas.
   b. Gas exploration and development have declined as costs, but not prices have risen.
   c. Higher prices would provide the industry with more incentive to explore and produce.
   d. Abrogation of contractually agreed prices by the FPC adds an uncertainty that the industry must regard as an additional risk factor.

RECOMMENDATIONS

1. Natural gas prices should be allowed to rise to a level sufficient to provide incentives for exploration, and for development of alternative sources.
2. A deadline or timetable should be imposed on the regulatory process so that the industry can be assured of a decision on a particular request within a specified time. Contracts, once approved within this time frame, should be immune to rollback or alteration by the regulators.
3. A program should be developed to allow exploration and development of resources offshore and on public lands in consonance with the public's energy needs as balanced against its environmental desires. Once designed, the policy should be permanent so as to provide the industry with a stable reference against which to plan.

   In addition a leasing policy should be devised so coal on public land will be available for gasification consistent with suitable environmental regulations. Such a policy should take into account water needs of such processes (see Coal, pp. 43-44) as well as site and pipeline requirements.
4. Prices, subsidies or incentives should be structured to encourage research and development of alternative sources such as coal gasification and LNG.

NOTES TO CHAPTER FOUR

[1] Senate Committee on Interior and Insular Affairs, Natural Gas Policy Issues and Options, staff analysis pursuant to S. Res. 45, #93-20, p. 1.

[2] Ibid., p. 31.

[3] Ibid., Table 3.1, p. 32.

[4] Council on Environmental Quality, OCS Oil and Gas: An Environmental Assessment.

[5] Senate Committee on Interior and Insular Affairs, op. cit.

[6] Ibid.

[7] Interstate Natural Gas Association.

[8] Council on Environmental Quality, op. cit.

[9] Ibid.

[10] Interstate Natural Gas Association.

Chapter Five

Coal

For the first half of the 20th Century, America's preeminent fuel was coal. Following World War II, however, the nation underwent a massive conversion, expanding its use of oil and natural gas at the expense of coal.

By the mid-1960s, coal was generally considered the fuel of the past. Dirty, dangerous to mine, expensive to transport, its share of the energy market seemed doomed to continued decline. In addition, its major market, electric utilities, seemed destined to be taken over by nuclear power.

But rapidly climbing international oil prices, delays in the development of nuclear power, the extent of U.S. coal reserves and their promise as a source of oil and gas have made coal relatively attractive again.

Coal's biggest asset is quantity. U.S. reserves are currently estimated at 750-800 billion tons,[1] about 50 percent of which is thought to be recoverable with current methods, enough to last several hundred years even at sharply increased consumption rates.

But the problems that originally cost coal its markets remain. They can be solved, but there will be costs. Coal can provide the energy, but it is up to the nation to decide the terms.

INDUSTRY BACKGROUND

Coal occurs across vast areas of the United States—in a belt along the Appalachian Mountains from Pennsylvania to Alabama, under large areas of Illinois, Indiana, Missouri and neighboring states, and in large deposits from Montana south to Arizona and New Mexico.

It is recovered through underground mining and through stripping. Underground mining involves tunneling along the coal seam, breaking up the coal and transporting it to the surface on hopper cars or conveyers. Stripping involves removal of the overlying material and scooping out the coal with power shovels or similar earth-moving equipment. U.S. coal production–602 million tons in 1973–is almost evenly divided between the two methods.

Coal from the East is noted for its high heat value and is widely used as boiler fuel and in metallurgical processing both domestically and abroad. Midwestern coal is used extensively by power companies in that area.

Much of the Eastern and Midwestern coal is high in sulfur, however, which limits use where strict air-pollution standards are in effect.

In the West, however, the coal, though relatively low in heat value, is also low in sulfur. Consequently, considerable attention has focused on these deposits, many of which are also close enough to the surface to be strip mined.

The need for increased coal production comes against the background of a troubled industry. From its dominant position, the fuel fell into relative disuse–the modern low point in coal consumption, 400 million tons, was reached in 1958–as oil and gas, their production expanded during the war, became widely available.

Big pipelines made petroleum products available at reduced prices; at the same time labor disputes disrupted coal production, accelerating the changeover. Despite a decision by John L. Lewis to embrace mechanization–which improved productivity sharply–consumption continued to decline.

During the 1950s, the leading producers were independent companies and "captive" mines owned by steel, utility and other companies to supply their own needs. With the coming of the 1960s, however, the outlook began to change. Giant corporations began acquiring leading coal producers–Consolidation Coal, 1973 production 60 million tons, was bought by Continental Oil; Island Creek Coal, 22 million tons, by Occidental Petroleum; Peabody Coal, 69 million tons, by Kennecott Copper; Ayrshire Collieries, 16 million tons, by American Metal Climax.

While critics have charged that such acquisitions, especially those involving large oil companies, would lead to energy "monopolies," the coal industry has benefited from an infusion of capital and expertise, particularly in research.

The ten largest coal companies are:

Company	*1973 Production (millions of tons)*	*Parent Company*
Peabody Coal Company	69.9	Kennecott Copper Corp.
Consolidation Coal Company	60.5	Continental Oil
Island Creek	22.9	Occidental Petroleum
Pittston Company	18.8	(independent)
Amax	16.7	American Metal Climax

Table continued

Company	*1973 Production (millions of tons)*	*Parent Company*
U.S. Steel	16.2	captive mines*
Bethlehem Steel	14.1	captive mines*
North American Coal Company	12.5	(independent)
Old Ben Coal	10.8	Standard Oil of Ohio
Eastern Associated Coal	10.6	Eastern Gas & Fuel Associates, Boston

*Output is for the use of the parent.

New sources of research support are particularly welcome in the coal industry, where profits have traditionally been low and funds available for research limited. Also, coal has suffered from competition from nuclear energy, much of its technology developed with the support of the government. Many utilities in need of additional capacity turned to atomic power at the expense of coal.

Even into the early 1960s coal production capacity exceeded demand to such an extent that operators were forced to sell coal at little or no profit.

The situation appears to be changing, and operators are struggling to increase production to meet demand. But profits are still not great; legislation has imposed stiff new requirements on the industry, and impending bills threaten more. There are great uncertainties regarding its production and use. The industry is emerging from a long bad time. If it is to expand as the Administration's Operation Independence envisions 962 million tons in 1980—it must be assured of strong national support.

COAL INDUSTRY ISSUES

Regulation

High on the list of concerns to the coal industry is the future of government regulation. The Coal Mine Safety Act of 1969 and impending strip mine regulations have imposed and threaten to further impose stiff burdens on the industry.

Safety. Productivity has declined since passage of the Coal Mine Safety Act. Many small, but collectively significant, mines have been forced to close. Costs for other mines have risen sharply.

It is difficult to determine how much the Coal Mine Safety Act will add to the overall cost of coal. Mines vary widely; some will find it necessary to buy large amounts of new equipment, and most will find it necessary to add more men. Whatever the basic cost of compliance—and that could easily add $1. a ton to a product whose average 1973 price at the mine mouth was $8.50

industry spokesmen say—there will be yet more costs resulting from the decline in productivity.

Strip Mining. Strip mine legislation is potentially even more important, given the potential of the Western coals. It is significant enough under present circumstances:

> The National Coal Association estimates that almost 715 billion kilowatt hours of electricity were produced by coal in 1971, 44 percent of the nation's electricity. To produce this the utilities consumed 332 million tons of coal; 202 million tons, or 60.8 percent, came from strip mines. Electric power generation accounts for 73 percent of all surface mined coal.[2]

There have been calls for an outright ban on surface mining. Those seeking such a ban or other measures that would significantly reduce stripping recommend the entire industry convert to deep mining.

The costs of such a switch, if it were deemed desirable, would be enormous. But more important, few in the industry believe such a change could be implemented. According to E. R. Phelps, president of Peabody Coal Co., "There is no possibility that alternative underground production in the amounts needed can be made available. The only option open to the utilities, assuming a severe cutback in surface production, would be foreign oil, a clearly unacceptable alternative for many reasons.

> . . . Furthermore, it is not realistic to expect that surfaced mined coal could be replaced by production from underground mines. While there are ample underground reserves, to produce the 276 million tons of surface coal mined last year [1972] would require 138 additional underground coal mines of two million tons annual capacity, which would require a capital investment of $3.5 billion, and three to five years before full production could be anticipated. The effect on the cost of coal would be tremendous. . . .[3]

In addition, other requirements have been proposed that would make strip mining much more expensive in some places and impossible in others. For instance a bill introduced in the House of Representatives in 1973 would have:

1. required strip operators to "segregate and preserve topsoil;"
2. returned strip-mined land to the approximate original contour;
3. prohibited piling of dirt on slopes greater than 14 degrees, and prohibited mining slopes greater than 20 degrees.

Coal operators pointed out in hearings on the bill that the first two requirements would greatly increase the expense of surface mining, and the third would eliminate it in much of the Appalachian region. Segregation of topsoil is particularly difficult in a large operation where overburden removed from mining in one place is used to reclaim already mined areas in another. To have to set the topsoil aside would impose a heavy burden on the operator without corresponding benefit in reclamation.

Returning of strip-mined land to its original contour might actually involve disturbance of more land than was involved in the mining. If 50 feet of overburden is removed from 50 feet of coal and the coal taken out, a requirement to even off the highwall would mean that additional material would have to be brought in or the highwall graded off to a distance many feet from the edge of the original mine.

Finally, the slope restrictions would simply eliminate many areas of West Virginia and neighboring states, where slopes exceed the specified angles.

The bill has not passed, but it illustrates the problems involved in regulating strip mining. The industry, says Carl E. Bagge, president of the National Coal Association, supports "a reasonable, realistic law to set federal guidelines for strip mining," but urges great care lest regulations cost the nation more than they add benefits.

Transportation

A major factor in the cost of coal is transportation. A Gulf Coast utility using Illinois coal transported by barge could expect to pay $12.50 a ton for a product whose mine mouth cost is about $6. An East Coast utility using low-sulfur Western coal transported by rail could expect transportation costs to run triple the product costs—$12 a ton to move $4 a ton coal.[4]

As the nation turns to low sulfur Western coal, as now seems likely, methods of reducing transportation cost take on added significance. One possibility is installation of coal-fired generators and coal gasification and liquefaction plants at or near the mines.

Water

Coal gasification and liquefaction both require large amounts of water. Unfortunately, most Western coal occurs in arid or semi-arid areas. Water can be made available and the utility industry has already done some preliminary planning.

Western water supplies are, however, in demand by other interests—agriculture, conservationists, developers—and allocation is proving a thorny problem that has yet to be worked out.

A 1971 plan by 25 utilities for development of coal and water resources in the Gillette-Colstrip area of Wyoming and Montana was attacked in the courts by suits disputing the Bureau of Reclamation's right to designate large amounts of water for the project.

The water problem is far from solved, and whatever its resolution, delays seem inevitable. A battle of the proportions of the one over the Alaska pipeline could develop, which could defer extensive development for years.

Research

Three major issues confront the industry in the research field: sulfur removal, economic production of synthetic fuels, and mining technology.

An important factor in the movement away from coal in the generation of electricity has been air pollution. The main pollutants are particulates, nitrogen oxides and sulfur oxides.

Sulfur oxides are the most important. Clean air standards can be met only by burning coal with one percent or less sulfur. A variety of stack scrubbing devices are now under development to remove sulfur oxides from exhaust gases at power plants. To date, no economical efficient process has been proven, though a number of approaches appear promising.

The importance of sulfur removal cannot be overstated. If 50 percent of the oil and gas now used in power generation could be replaced by coal, 200 million barrels of oil and 1.9 trillion cubic feet of gas[5] could be saved each year. Further, U.S. dependence on foreign oil would be significantly reduced.

Technology exists now for converting coal into gas and liquid fuels such as gasoline and methanol. Some of the processes date back to the mid-19th Century. However, until the energy crisis, most were of little interest because of the availability of cheap oil and natural gas.

Now, rapidly rising prices for imported oil have generated new interest in such synthetics.

Current processes are not as efficient as might be desired. Conversion of coal gas to methane for use with natural gas or to methanol or liquid hydrocarbons involves usable energy losses with each conversion step.

Research is needed into these problems, as well as into the questions of sulfur removal—use of low-sulfur anthracite avoids this problem, but supplies are limited and the industry currently moribund.

Large-scale investments are needed, but the resources of the coal industry are limited, so that some form of outside assistance may be necessary until the competitiveness of the various processes is established.

At current and still uncertain estimates, the costs of coal gasification are put at about $1.25-1.50 per million B.T.U.; thus a 250 million-cubic-foot-per-day plant would cost about $400-500 million. Syncrude costs range in the same order of magnitude.

From the early 1950s to the late 1960s, coal-mine productivity showed steady gains. Development of continuous mining equipment in underground mines and of gigantic earth-moving equipment for surface work pushed productivity steadily forward.

Then in 1969 a sharp falloff began. With the successful development of machinery, research emphasis had been switched to end-use rather than mining technology. Now the industry finds itself playing catch-up, and with less outside help here than in other areas. While companies in other fields are investigating coal uses, such as gasification, there is not enough work on mining itself.

Thus, unless headed off by research, the coal industry could be headed into a period of declining productivity at a time when the nation can ill afford it.

Labor

An unknown factor in the future of the coal industry is labor. Labor relations in the coal fields have traditionally been poor, and today the United Mine Workers union is undergoing a change of leadership and orientation.

The new reform leadership under union president Arnold Miller, negotiated its first round of new contracts in 1974, and may face a problem of elevated expectations among the members. However, industry officials seem optimistic that their labor problems can be solved.

Other labor considerations include the shortage of competent engineering personnel and the bi-polar age distribution of the work force.

CONCLUSIONS

1. Compliance with federal safety requirements presents a major difficulty for the industry in a period of declining productivity and increasing demand for output.
   a. Costs of compliance will be passed on to consumers, and they are likely to be large.
   b. Production will be ended at small, marginally economical mines, and there will be production losses at larger mines.
2. Strip-mine output will continue to grow in importance unless curbed by legislation.
   a. Low-sulfur Western coals will increase in use, and they are most efficiently recovered by stripping.
   b. One-third of the nation's electricity is now produced from surface-mined coal. The quantities involved are too large to be replaced by underground mining in the near future and probably in practical terms, too large ever to be replaced by deep mines.
   c. Some of the proposed restrictions on strip mining would eliminate large reserves from production. Others would vastly increase the cost of strip-mined coal.
   d. Restrictions on strip mining could cut the nation off from its largest readily available sources of additional energy.
3. Pollution is a major limiting factor on coal use.
   a. Sulfur removal, either from the coal itself or from stack gases, has not yet proven commercially feasible. Several methods are being tested, and it is hoped one or more will prove successful.

   b. Failure to develop adequate sulfur removal, coupled with continued stringent air-pollution standards, would prevent use of coal to the extent necessary to alleviate increased dependence on low-sulfur foreign oil and the concommitant uncertainties of supply.
   c. Adherence to the standards of the Clean Air Act of 1970 and its requirement that all sulfur dioxide emissions be in conformity by 1975-76 could preclude the use of, according to some estimates, 40 percent of the steam coal currently in use, which would jeopardize the output capability of such major systems as the Tennessee Valley Authority and American Electric Power Company:
4. Full utilization of low-sulfur Western coal is linked to the availability of water needed in gasification and liquefaction processes.
   a. Most of the Western coal occurs in chronically water-short areas. Supplies can be made available, but they are likely to be sought by competing interests, such as agriculture, and their use is likely to be considered a threat to the environment by conservationists.
   b. Nonavailability of water would compel transportation of Western coal long distances at great cost—a cost that consumers would be required to pay.
5. Coal is the potential source of additional supplies of gas and liquid fuels.
   a. Processes for conversion exist, but require further development. Also, installation of conversion plants requires large sums, which will most likely be supplied from outside the industry or from larger companies with substantial coal subsidiaries.
   b. Fluctuations in foreign crude oil prices create uncertainties about future competitiveness of synthetics. Foreign producers are in a position to produce a yo-yo effect on prices, which can only be offset if industry, government and the public are willing to resist it.
   c. Research, particularly into more efficient processes for creating liquid fuels, could lead to advances that would make these fuels more competitive.
6. Labor presents some uncertainty for the immediate future, but is not considered by management to be an intractable long term problem.

RECOMMENDATIONS

1. Maximum flexibility should be built into federal safety standards, and procedures for case-by-case review should be implemented so that no mine that might reasonably be kept open is forced to close.
2. Strip mine regulations should be free of rigid, across-the-board restrictions that impose unnecessary hardship on some operators, and any restriction likely to reduce strip mine output should be carefully weighed against the nation's need for that output.

3. Priority should be given to energy-producers' water needs in developing Western coal. In addition, acquisition of leases and plant sites for exploiting reserves on public lands should be facilitated.
4. Government support, in terms of direct subsidies, tax or price aid, should be given to research in sulfur removal, and, in the meantime, air-pollution standards should be reconsidered on a case-by-case basis to allow utilities to use high-sulfur coal to prevent economic hardship and/or risks of supply interruptions.
5. The government should similarly support research into better methods of converting coal to other fuels, and incentives—tax advantages, guaranteed loans or grants—should be extended to companies to offset the uncertainties involved in large-scale investment in conversion facilities.

NOTES TO CHAPTER FIVE

[1] National Coal Association.

[2] Statement of E. R. Phelps, president, Peabody Coal Company, at hearings on HR 3, 1973, p. 1433.

[3] Ibid.

[4] The World Crude Oil Outlook, a report prepared by Ronald D. Lewison for William D. Witter, Inc., 1974, p. 29.

[5] Dunham, James T. et al, *High-Sulfur Coal for Generating Electricity*, Science, 19 April 74, pp. 346-351.

Chapter Six

Conclusions

ENERGY POLICY

The major tasks facing the energy industry over the next two decades are, first, meeting the energy demands of the nation, and, second, coping with rapidly changing circumstances that introduce major uncertainties and difficulties into programs designed to assure adequate energy supplies.

Traditional factors such as return on investment, maintaining corporate viability and serving competitive or franchised markets continue to play their dominant role. However, the industry has moved away from a position of being able to meet demand with relatively low prices, domestic self-sufficience, marketplace pricing, and public policies focused primarily on the assurance of adequate energy supplies. Environmental and health and safety regulations, lack of access to resources, restrictive economic regulations and anti-energy industry attitudes have suddenly combined to make assurance of energy supplies an unusually formidable task.

Public, commercial and industrial demand for all forms of energy is expected to increase substantially, and in the case of oil and gas, is already beyond current domestic supply capability. Demand is expected to increase despite sharp price increases brought on by major new environmental and safety regulations, general economic conditions and increased costs of exploiting marginal reserves. Major dependence on imports adds geopolitical and monetary problems of unprecedented magnitude.

The now strongly interlocked government, industry and public decision-making system is deficient in its operation and makes industry decision-making about future energy supply most difficult. In order to develop a workable system, the nation requires a broadly shared understanding of: (1) the nation's energy demands; (2) what steps are necessary by industry and government; and (3) what the costs are to meet the demands for energy and environmental health and safety and independence goals.

To evolve such an understanding and national policies based on it, a suitable forum is needed in which the requirements of energy users and suppliers can be examined coolly and rationally, and national goals and policies can be developed on an informed basis. Currently, no such forum exists and it can be argued that the lack of one has contributed substantially to the national confusion surrounding the continuing energy crisis.

It is recommended, therefore, that the President draw together appropriately qualified representatives of industry, government, and the scientific community to form a national Energy Policy Council reporting directly to the President. It is further recommended that, while this Council would be largely supported by federal funds, its function should be that of a non-political, independent and permanent body to study and advised on the nation's energy supply and demand and conflicting national goals—and to recommend appropriate policies and legislation for the orderly achievement of those goals. The Council would be advisory but authoritative.

Among the Council's most immediate tasks would be a re-examination of economic incentives and disincentives which impact upon the oil, gas, coal, nuclear fuel and electric utility industries. It would make recommendations for change in the light of national policy alternatives. Major new federal funding of research and development should be coordinated with the council's policy recommendations so as to guide research into areas that properly reflect evolving national goals and policies.

The Council also could be expected to play a major role in the coordination of governmental policy development and decision-making with appropriate and responsible industry and public groups. It could additionally recommend to the various levels of government more effective means for the coordination of their own decision-making and administrative machinery in the energy area, including the coordination of regulatory criteria and actions with national policies and the public's demand for energy.

PRICES AND PROFITS

Government regulation, particularly in the area of prices, is a major inhibiting factor in the production of energy. Electric utilities are having difficulty raising capital for new construction; natural gas producers need more incentive to explore for new reserves; price controls in the petroleum industry lead to dislocations in the economics of refining; coal companies have signed long-term contracts only to have the agreed-upon price disallowed.

Removal or relaxation of these restrictions would do much to improve the supply situation. Industry could set prices to reflect costs plus a reasonable profit. The incentives would be clear. Exploration would be revitalized—as the upsurge in drilling that has accompanied the 1973-74 price rises demonstrates. Higher prices would also bring more sources into the realm of commercial feasibility, notably synthetic gas and synthetic crude oil.

Increased supplies should also provide a natural brake on prices. After initial rises to catch up with costs, prices may begin to reflect the presence of new supplies and increases may be curbed.

Profits must be adequate for American industry to function and to attract the vast financial resources required to provide energy. When profitability is threatened or restricted, not only incentive is lost, but economic viability as well. In recent months some companies have recorded what appear to be vast profits. Indeed the sums are large. But the margin on any particular product is small. If the energy industry is to do the job of supplying energy to meet U.S. demands, properly developed economic incentives will get the job done, and in a manner which satisfies the nation's priorities.

RATE REGULATION PROCEDURES

In an era of rapid change, regulatory procedures, never particularly flexible, have become a millstone around the necks of the industries they were designed to foster. This is especially true of the electric utility industry.

To improve this situation, it is recommended that the present rate-setting process be somewhat modified. The basic standard would remain the reasonable rate of return as approved by the regulatory agencies. However, instead of requiring the utility to request and justify rate increases and then wait to have them approved, the utility would be allowed to put new rates into effect as required to maintain the approved rate of return, and then justify them to the regulatory agency. Under such a system, utilities would be able to generate revenues and returns as allowed by law, with the proviso that they be subject to price rollbacks and penalties should they be found to have implemented unjustified rate increases.

Such a reversal could be implemented without major changes in the present regulatory structure. It would eliminate the severe problem companies face now of rates that cannot keep up with costs because costs are increasing so rapidly that whatever data they supply to the regulators is obsolete by the time a decision is made. Strictly applied, penalty provisions should be sufficient to prevent abuse.

This simple change could do much to protect the utility industry and its customers from the inevitable negative effects of a financially weakened industry.

ENVIRONMENTAL POLICIES

Few in any segment of American life today would argue with attempting to improve the environment. But after less than half a decade of "environmental consciousness," it would be fool-hardy to assert that the best and most effective methods for achieving it have been devised.

On the contrary, some of the methods and timetables called for in legislation now on the books are clearly unworkable and even counterproductive. These should be reassessed.

Three areas particularly need examination. First, attempts to impose technical "fixes" that have not been proven either as to effectiveness or to cost. Initial installation and/or retrofitting of pollution control devices can involve enormous cost. Most companies are willing to bear these costs if results are assured, but few are anxious to install devices that later prove inadequate and which then require replacement with still other devices.

The second area is the cost and benefit relationship. It is frequently possible to eliminate a given percentage of a certain pollutant at a bearable cost. But then to eliminate more requires vastly greater costs. Frequently, environmental regulations call for removal of so much of a pollutant that costs become completely disproportionate to the benefit.

Third, standards for specific pollutants should be continually reassessed so that they can be tightened or loosened on the basis of the best relevant findings.

Similarly, restrictions that remove some resources from possible production altogether should be reconsidered. The proposed ban on strip mining on slopes greater than 14 degrees is an example. This ban would prevent strip mining in many areas—but is this elimination of large quantities of cheap, easily available coal desirable? Strip mined areas can be reclaimed. They may not be restored to their exact condition before the mining, but neither need they be the eyesores and environmental hazards some assert.

STRUCTURE OF THE ELECTRIC UTILITY INDUSTRY

The electric utility industry is playing an ever more important role in supplying the nation with energy. It faces even larger tasks ahead.

In 1935, the Public Utility Holding Company Act, under Section 30, directed the Securities and Exchange Commission to make studies of the industry for the purpose of recommending the "type and size of geographically and economically integrated public utility systems, which, having regard for the nature and character of the locality served, can best promote and harmonize the interests of the public, the investor, and the consumer." Further, the Act expressly recognizes that the appropriate size of integrated systems is to be determined in light of the current state of the art and contemplates increasingly larger systems as warranted by advancing technology.

It is recommended that the SEC promptly organize for and make the studies referred to.

ANTITRUST RESTRICTIONS

In an era of acute cost-consciousness in energy, opportunities may exist for considerable savings through pooled resources and joint ventures.

The petroleum industry, which has long engaged in joint ventures because of the high costs and risks involved, currently faces the possibility of antitrust action to prohibit the practice at least where more than one large company is involved.

The regulated and non-regulated energy companies have clearly separable financial and operational interests. The non-regulated companies—oil, coal, nuclear—compete in the marketplace in pursuit of their separable interests. They also collaborate where their separable interests are not impaired, but rather enhanced, and where antitrust law allows.

The magnitude and complexity of the energy supply problem requires the efficient use of all institutions which can contribute to solutions. To the extent that collaboration by energy industry companies either expedites or enhances the economics of resource development and supply or building supply facilities, or conduct of research and development, such collaboration should be encouraged rather than discouraged by law.

It is recommended, therefore, that the application of anti-trust laws to activities of the energy industry which are critical to the assurance of adequate energy supplies be examined and modified on a case-by-case basis to allow collaboration where prospective supply benefits are demonstrable and in the public interest.

Appendixes

Appendixes

During the course of research for this report, factors were identified that influence significantly decision-making processes within the energy industry. The study was also to help determine how a national energy policy can be fashioned so that private industry will have both the incentive and capability to carry it out.

Appendix A provides an outline of the decision-making factors involved for the industry as a whole. Appendix B provides an outline of factors for fashioning national energy policy. Appendixes C through F focus in similar fashion on the oil, electric utility, gas and coal industries individually.

The factors influencing industry decision-making are categorized as they are (see I.A., I.B., I.C., II.A. etc.) to help clarify the sources of the influences. The categories are intended to illustrate that many of the influences are fundamental to the market system and self-determination on the part of independent private enterprises; some are established and controlled by governments; and that the money market and public attitudes are major independent factors. The central thrust of any energy policy which presumes the continuation of energy supply by the private sector must assure that the interaction of all of the influences is harmonized. Listing factors in this manner, while it helps to define the influences, obviously cannot depict the dynamics and variety of industry decision-making and its responsiveness to the wide variety of social and political forces exemplified by some of the influences listed.

The remaining appendixes provide amplification and additional detail on certain aspects of a few of the problems addressed in the body of the report.

Appendix A

Factors Influencing Energy Industry Decisions

I. Decision-making in the investor-owned, nonregulated[a] industries–oil, coal, nuclear, and gas production–is influenced principally by:

A. Factors established by or under the control of the industry:

1. Return on investment objectives established independently by each company.
2. Each company's perception of its public service obligations.
3. Each company's perception of its other objectives and obligations. (These may include or be based on individual companies' perception of, and/or attempt to meet, societal demands not yet internalized.)
4. Perceived and actual ability of individual companies and industry groups to influence the enactment and administration of pertinent government policy, law and regulation.
5. Perceptions by each company of the demand(s) and opportunities in their domestic and/or foreign markets and all factors which pertain to optimizing short- and long-term return on investment.
6. Resources controllable by each company–physical, financial, management, labor, technological.
7. Ability to select those opportunities available to each company

[a]As identified here, non-regulated industries have no legally mandated service obligation and normally no regulation of the prices charged by them. Certain activities of the oil, coal, nuclear and gas producing industries are influenced markedly by regulations but they do not meet the foregoing test. *Regulated* industries are those that have a combination of legally imposed public utility service obligations, and a limit on rate of return on investment.

perceived as most likely to result in meeting its short- and long-term return on investment objectives.

8. Ability to select any other objectives or assumption of any obligations by each company which may be achievable without conflicting with return on investment objectives, or which are perceived likely to support meeting investment objectives.
9. Perceptions by each company of the suitability of technological innovations for commercial application and avenues of possible future technological development.
10. Lead time (the time, generally very long, required to plan, develop physical resources, and build facilities to meet current and future demand.)
11. Perceptions that conservation of energy is requisite to help alleviate critical near- and long-term supply problems.

B. Factors established by or under the control of government:

1. Federal and state government policy, law and regulation pertaining to each company's operations, and how these are implemented and interpreted by the courts. (Laws and regulation governing environmental, health and safety matters, prices and taxes, currently have major influence.)
2. Control by others—such as the U.S. and foreign governments—of access to basic fuel resources.
3. Availability of land (and ocean areas) for building supply facilities, rights-of-way, and for mining energy resources.

C. Factors established by or under the control of others:

1. Perceived and actual availability of capital on terms dictated by the money market.
2. Public attitudes, perceived to be negative, and interacting with the executive, legislative and judicial processes, which impair individual companies achieving financial and/or operation objectives.

II. Decision-making in the investor-owned, regulated industries—electric utilities, gas transmission and distribution—is influenced principally by:

A. Factors established by or under the control of the industry:

1. Each company's perception of its own interests and obligations. These may include or be based on individual company's perceptions of and/or attempt to meet societal demands.
2. Perceived and actual ability of individual companies and industry groups to influence the enactment and administration of pertinent government policy, law and regulation.
3. Perception by individual companies of the demand(s) and opportunities in their domestic markets.

4. Resources controllable by each company–physical, financial, managament, labor, technological. Some physical resources, such as fuels for the electric utility industry, and gas supply for the gas transmission and distribution industry, are now critical because of supply shortages and/or government constraints.
5. The perception of the electric utility industry that, long-term, the capabilities and resources of the industry, and those under the sovereign control of the U.S. are adequate for it to meet its obligations and objectives (increased electrification of the U.S.), with rationally derived and administered national policy, law and regulation.
6. The perception of the gas transmission and distribution industry that, while the physical resource base under the control of the U.S. may be large, there are substantial uncertainties about future supply and price.
7. Desire of the private sector to work effectively, openly, and in harmony with federal and state governments and with the public.
8. Lead time (the time, generally very long, required to plan and build facilities to catch up with current demand and meet future demand.)
9. Perceptions that conservation of energy is requisite to alleviate critical present and future supply problems.

B. Factors established by or under the control of government:
1. Each company's public utility obligations derived from government franchises.
2. Return on investment objectives established by each company and approved by appropriate government regulatory agencies.
3. Federal and state government policy, law and regulation pertaining to each company's operations, and how these are administered and interpreted by the courts. (Laws and regulations governing environmental, health and safety matters, and prices, currently have major influence.)
4. Availability of sites for generation and other facilities, and rights-of-way for transmission.

C. Factors established by or under the control of others:
1. Perceived and actual availability of capital on terms dictated by the money market. Also restrictions, such as "coverage" requirements, which limit companies from availing themselves of capital.
2. Public attitudes, perceived to be troublesome, which impair individual companies' ability to achieve operational and financial objectives necessary for the companies to fulfill their legally mandated local, regional and national obligations.

III. Decision-making in the government-owned electric utility industry is influenced principally by:

A. Essentially all of the same factors as those for the investor-owned utility industry, except those related to financing and taxation, and, to some extent, public attitudes. These are discussed below.

B. Factors established by or under the control of the utility:

   1. Revenue objectives, although variable with each entity, are not understood to be based on return on investment criteria. They are not as uniformly influenced, as are those of the investor-owned utilities, by the views of the money market as to the adequacy of financial performance. (TVA, an exception to this, goes to the same money market as the investor-owned industry. In lieu of earnings objectives, TVA has revenue reserve objectives (tests) which, if met, assure favorable money availability and costs. TVA is able to set its own rates at levels which assure sufficient revenues to meet its test objectives.) The bond market's support of municipal electric utilities is generally based on a pledge of the revenues of the utility which are controllable by the municipality.
   2. Substantial coordination and integration with, and dependence on, the private sector of the industry.
   3. A perception that public attitudes are not as negative as those generally accorded the private industry, coupled with a perception of being more a part of the body politic.

C. Factors established by or under the control of government:

   1. Federal government executive and legislative attitudes, and appropriations governing availability of capital funds for expansion of government systems such as Bonneville Power Administration and Southeast Power Administration.
   2. Generally more favorable tax treatment and money costs then the private industry. (It should be noted that privately owned utilities have the advantages of the investment tax credit, liberalized depreciation, accelerated amortization, and other tax benefits.)

D. Factors established by or under the control of others:

   1. Public opinion as in B.3 above.

IV. Some General Conclusions On Factors Influencing Decisions

A. What decision making has been:

The perceived mission of the U.S. energy industry is to meet the demands of the U.S. (and in some instances other nations) for all forms of energy at the lowest prices consistent with a reasonable profit. For the investor-owned regulated electric and gas utility industries this is mandated by law. Individual companies and industries are expected to advance their own interests and to worry about the interest of others as they are required to, or think they should. Decisions are made to that end. This has been done traditionally within the constraints, or with the encouragement of, public policy, law, and regulation which allowed the energy industry to go about its work and make enough money to survive, grow, and sometimes prosper. Decision-making was, until recently, a relatively orderly matter for the energy industry. Industry was able to work relatively harmoniously with government. The result was low-cost energy and, for the most part, the consuming public was contented.

B. Decision-making and changed circumstances:

Industry decision making has changed. It is now much more difficult. Basic influences–such as supply and demand, service obligations, return on investment, law, regulation and "enlightened self-interest" continue to influence industry decision-making. But the character and impact of these influences, and industry's ability to respond to them, have changed dramatically, due to factors beyond the control of industry.

Demand for all forms of energy increases. The U.S. resource base, especially the "lower 48," ends up short on natural gas and domestic oil and prospectively long on coal and nuclear. It becomes clear that sovereign nations controlling oil reserves needed by the U.S. will supply oil based on their own perception of their own self-interest–not America's.

Congress enacts environmental legislation impacting heavily and negatively on production of coal, nuclear, oil, gas and electricity. The environmental legislation brings with it an invitation for some of the public to intervene in new and cumbersome government/industry decision-making–dealing with the extent, character, location, and operation of energy supply facilities.

Government control of energy prices dominates the most critical factors in energy industry decision making–responsibility to meet public demands and necessity to receive an adequate return on investment.

Other government decisions controlling now critical supply do not get made and industry cannot develop needed resources.

The combination of a deficient, interlocked government/industry decision-making process and public intervention results in major delays and increased costs in operating and expanding the energy supply system, where the tolerance for delays and price increases is very limited.

C. Decision-making today:

The energy industry today cannot make and implement decisions about the extent, character and cost of energy supply systems; nor, in the instance of the regulated industries, price its products at levels which produce revenues sufficient to assure its long-term financial viability and supply capability. Layered on top of the industry decision-making process is a vast federal-state government decision-making apparatus which has very broad authority governing details of supply and pricing. There is a classic division of authority, responsibility and accountability between government and industry.

The government decision-making process generally does not yet work in concert with the industry decision-making process. The public, not sympathetic to the industry and its problems, and perhaps unaware of the adverse impact upon themselves, makes decision making more difficult. The cumulative effect measurably impairs industry's ability to solve supply problems very difficult to begin with.

Appendix B

Factors in Formulating a National Energy Policy

One purpose of this study was to help determine how a national energy policy can be fashioned so that private industry will have both the incentive and capability to effectively carry it out.

A national energy policy can be fashioned so that private industry will have both the incentive and capability to carry it out if the policy is based on recognition of the following:

A. Planning and economic factors:
1. Private industry should be required to contribute to the national process of fashioning national goals, and policies and standards, to assure that they are based on:
    a. the realities of the supply and demand problems.
    b. understanding the feasibility and cost of attaining environmental and other goals.
    c. a realistic perception by policy makers of what is involved in balancing energy supply and environmental goals.
2. Industry has a long-term vested interest in conforming to societal dictates, and great incentive to participate in forging and carrying out national policies deemed by the public to be in the public interest.
3. Private industry must be responsive to money in return for goods and services, and must have something left over for profit.
4. The U.S. energy industry has financial, managerial, labor and technical capabilities which, in quantity and quality, are unsurpassed by those of any other nation, and can implement any policies which are rationally derived, presuming manageable access to energy raw materials.
5. Industry's capabilities will support any combination of research and development and supply programs based on rationally derived demand, supply, environmental, economic and political goals. (Presuming a need to increase energy supply, additional supplies of engineering and

skilled construction labor are required by the energy industry. Also, the adequacy of the financial resources depends singularly on the money market's perceiving that energy industry companies are able to make a reasonable return.)

6. Private industry responds willingly to rationally derived societal dictates which do not impair its ability to function or to make reasonable return on investment.
7. Private industry has the capability (and strong desire) to participate in a reasoned public dialogue on all of the issues and options involved in creating and implementing energy policies and in providing energy supplies.
8. Within private industry there are different, company by company, perceptions of what financial or other incentives are appropriate for any given undertaking—and it is easy to find out what they are.

B. Objectives:

1. In the investor-owned regulated gas and electric utility industries, and the government(s)-owned electric utility industry, the obligation to meet customer demand is a strong influence. This is coupled with a perception that meeting demand is of long-term public benefit.
2. Private industry has strong incentives, based on enlightened self-interest, to satisfy societal dictates and to work in a harmonious and orderly fashion with the federal and state governments and the public.

C. Regulation:

1. The capabilities of U.S. private industry can be efficiently used only if law and regulation governing it are enacted, administered and adjudicated in a discerning and orderly manner. This manner must be consistent with effective and timely industry management and execution of planning, financing, design, construction, operation, and research and development of energy supply programs.
2. Stability in policy, law, regulation, taxes and any other governmental influences on industry decision-making is necessary for industry to be able to plan and carry out major supply programs.
3. Private industry has the capability to conform to any and all environmental, and health and safety, regulations and standards which do not substantially impair its ability to function or threaten its financial viability. Full use of this capability requires careful development and orderly introduction of any new standards and regulations and recognition by government and the public that consequent costs incurred must be reflected in prices.
4. Private industry also has the capability of questioning standards and regulations, measuring them against nationally agreed upon societal goals, and seeking to have them changed, if they are perceived to be without merit.

5. Federal and state land use and/or leasing policies, must permit industry access to energy resources, building sites, and rights-of-way on an orderly and timely basis.

D. "Internalizing" of externalities:

1. This concept needs to be better understood and carefully applied to specific industries, projects and products. (This is a mechanism which allows and requires the harmonizing of industry's and society's interests. Societal dictates become quantifiable industry objectives and obligations, and industry can perceive and capture all of the costs and benefits involved in acting to achieve them.)

General Observations on Formulating a National Energy Policy

Some perceive the U.S. energy problem as a failure of the uncontrollable and mighty energy industries to respond to public needs.

Some perceive the problem to result from:

- continued increase in demand consistent with U.S. economic and political principles now in force;
- new social dictates—especially environmental;
- new major dependence on non-U.S. supply for crude oil and products;
- deficient government organization and decision-making;
- public distrust or skepticism of industry and government.

Others perceive the problem as being an irrational and unnecessary public demand for all forms of energy in the face of major supply problems, industry deficiencies, and new environmental imperatives.

The basic problem is lack of a shared understanding—by government, industry and the public of what the issues and options are, and what actions should be taken. Such a shared understanding could endorse energy policies and a timely, constructive decision-making process which recognize:

- the actions and incentives required for industry to supply, build and operate the energy supply systems required to meet public demands;
- the skill and forbearance with which societal dictates must be imposed on the energy demand-supply systems without causing disruptions of supply.

A shared understanding requires an open, honest, informed, orderly dialogue. Such a dialogue and timely, constructive decisions being made in the simplest socially acceptable manner, though recognizably difficult to bring about, are essential.

While a shared understanding involves all interested parties, fewer, rather than more, decision makers are required. The need is for the fewest decisions being made by the fewest number of competent, responsible people, especially in the government/industry/public interactive system.

To the extent that government oversees or regulates the energy industries, its attitudes and actions must be constructive—providing for timely decision-making and necessary and acceptable action by industry. To the extent that government and industry decision-making has become intertwined on supply decisions their interaction must be constructive and coherent—or the supply of energy to the public is threatened. To the extent that the public is interested in industry decision-making, it must in the main rely on industry to satisfy demands for energy and on government to represent it in the continuing determination of how industry is to carry out its supply activities in a socially acceptable manner. The public has the long-term power to influence completely both government and industry.

The greatest possible number of decisions should be relegated to the smallest units that can competently make them. These are individual competent companies and individual informed consumers operating within the ground rules of law, regulation and the market place.

Along with the above is the need for greater public understanding of what's going on—what is being done for them, and what threatens their interests. The public can change or influence the decision makers when the latter do not conform to the public's perception of what the public interest really is.

Appendix C

Influences on Decision-making in the Oil Industry

I. Decision-making in the oil industry is influenced principally by:

A. Factors established by or under the control of the industry:

1. Each company's perception of its own objectives and obligations—principally return on investment.
2. Each company's perception of its public service obligations.
3. Each company's use of its own return on investment criteria which in turn dictates its investment, supply and pricing decisions.
4. Each company's perception of demand/supply and economic and political factors in areas pertinent to its interests.
5. Sophisticated analysis and planning capability relevant to worldwide and U.S. energy demand and supply.
6. Ability of individual companies to perceive and pursue opportunities in or out of the energy field beyond their historic interests, and availability of the resources requisite for same.
7. Resources—financial, management, labor and technological—deemed sufficient to correct supply deficiencies if economic incentives, law and regulation permit.
8. The perception that U.S. energy demands ultimately will be supplied from U.S.-controlled resources, but that continued long-term access to other resources is necessary. Current or near-term economics and/or law and regulation (especially governing environment requirements and leasing) now mitigate against rapid progress to that end.
9. Competition for supply and for markets; cooperation—within the constraints of federal law—when individual companies perceive that it is in their own separable interests to cooperate.

B. Factors established by or under the control of government:
   1. Uncertainties on quantity and price of supply deriving from the actions of foreign sovereign nations. Increased involvement of U.S. government in assuring supply from foreign nations and uncertainty as to the quality of the involvement.
   2. Uncertainties on quantity and price of supply deriving from the actions of the U.S. government, especially leasing and environmental regulations.
   3. Federal laws and rulings governing taxation.
   4. Environmental law and regulation at the federal and state level perceived to be restrictive and sometimes prohibitive.
   5. Uncertainties deriving from actions or inactions of U.S. federal and state executive, legislative and judicial processes which broadly influence decisions on investments, in, and installation of, supply systems and also influence costs, prices and profits.
   6. The perception that U.S. energy demands ultimately will be supplied principally from U.S.-controlled resources but that continued long-term access to other resources is necessary. Current or near-term economics and/or law and regulation (especially governing environmental and leasing) now mitigating against rapid progress to that end.

C. Factors established by or under the control of others:
   1. Current and projected supply difficulties, with domestic supply diminishing and major dependence on foreign sources being mandatory for an extended period of time.
   2. Rapidly increasing U.S. and worldwide demand for the industry's products and services.
   3. Public attitudes, perceived to be negative, interacting with the legislative and judicial process at the federal and state level—and impairing the ability of the individual companies to achieve their financial and operating objectives.

D. See also general conclusions on decision-making in the energy industry, Appendix A above.

II. Observations on energy policy and the oil industry:

Government policy, law and regulation should be enacted and administered to provide for:

1. Economic factors.
   a. Revenues perceived by the industry and the money market to be adequate, with reasonableness of rate of return being implicit.
   b. Encouraging and allowing prices for petroleum products to be at levels which provide the return on investment to assure the inflow of equity capital or access to debt capital required for expansion of oil industry supply systems. If, as a matter of

equity, or for any reason, adjustments are made in oil production tax incentives, prices should be allowed to adjust to levels commensurate with reasonable return on investment as perceived by the individual companies and the money market.

c. U.S. oil companies having economic incentives to produce and supply crude oil in the world market in conformance with U.S. government derived perceptions of U.S. interests.

2. Relations with foreign governments.

   a. The U.S. government having a stronger role, or presence, in assuring to the extent possible, that maximum practicable political stability and economic reasonableness are associated with provision of crude and products from foreign sources. Closer government/industry rapport and cooperation.

3. Supply and processing of crude oil.

   a. The firming up of crude and syncrude supply and the installation of supply systems necessary to meet the perceived public demand for oil products (and gas).

   b. Further exploration and development of domestic reserves of oil and gas by timely leasing of offshore and frontier areas. In doing this, applying environmental law and regulation to permit rational programs by industry.

   c. Importing crude and products consistent with meeting current and long-term demand, with full recognition of the control by others of the political and economic forces which influence availability and price of imports.

   d. The major supply systems required to import and process foreign (and where relevant domestic) crude–superports, supertankers, refineries, pipelines, distribution facilities . . . In doing this, applying environmental law and regulation to permit timely installation.

4. Supply of synthetics.

   a. A U.S. synthetic hydrocarbon fuels supply base as necessary to provide options for price leverage on, and ability to be independent of, foreign oil sources.

   b. Development and production of oil from shale and/or coal by providing, as necessary, special ad hoc price or tax incentives which result in a return on investment perceived sufficient by individual companies with relevant interests and capabilities; and administering leasing policies and environmental regulations to permit rational industrial programs.

   c. Development and production activities in coal mining, gasification, stack gas cleanup and other programs for the safe, environmentally acceptable and efficient mining and use of coal by providing, as necessary, special ad hoc price or tax incentives which result in a return on investment perceived sufficient by individual companies with relevant interests and capabilities; and administering leasing policies and environmental regulations to permit rational industrial programs.

   d. A program to assist in providing water requisite for synthetic fuels development and production programs, consistent with a balanced allocation of water to other important needs.

5. Competition.
   a. The oil industry being competitive, without being subject to harassment. Cooperation among companies—when a public benefit can be perceived such as cooperation on research and development or negotiating with foreign governments.
6. Regulation.
   a. Encouraging the oil industry to develop plans for distribution and marketing of fuel oil and heating oil and gasoline in the U.S. when supply is short—recognizing the public utility nature these products have assumed; the need for assuring supply of fuel and home heating oil at reasonable prices; and public demand for gasoline supplied reliably and at reasonable prices.
   b. Encouraging the oil industry to develop public-interest-oriented policies and plans for such supplies (in a. above) and to seek changes in policy, law and regulation if required to implement such policies and plans.
7. See also general observations on energy policy, Appendix B above.

Appendix D

Influences on Decision-making in the Electric Utility Industry

I. Decision-making in the electric utility industry is influenced principally by:

A. Factors established by or under the control of the industry:

1. A perceived continuing increase in demand for electricity, with this perception being shared by companies of the electric utility industry, other energy industries and the government—and confirmed by demands of industrial, commercial and residential customers. New uncertainty as to the extent of the growth in demand.
2. The perception that the use of electricity[a] relative to other energy forms such as gas and oil will and should for economic and other reasons increase significantly, with electricity eventually providing the majority of the energy consumed by industrial, commercial and residential consumers and significantly more of that required for transportation and other markets.
3. Ability of the internally non-competitive, investor-owned industry to exchange policy, management, planning, research, development, design, construction and operations information.
4. Ability of the private industry to organize and utilize in a substantive and broad manner industry-wide associations, such as EEI, which exchange technical, operating, marketing and other information; carry out industry-wide programs; and coordinate and carry forth industry views on a broad range of matters involving government and public interactions.
5. Ability of the public sector to organize and utilize in a substantive

[a] Generated principally from domestically available coal and nuclear fuels.

and broad manner an industry-wide association such as APPA which carries out many of the same functions as in 4. above.

6. Ability of the private and public sectors to work for the most part in harmony on planning, pooling, transmission and inter-ties, and especially on research and development, through the Electric Research Council and the newly formed Electric Power Research Institute.
7. General acceptance by the private and public sectors of the status quo.
8. General acceptance by companies within the private industry of the status quo as to the number and size of electric utility companies. This is buttressed by perceptions of antitrust law and the Holding Company Act mitigating against the orderly development of larger private systems, which are perceived by some within the private industry as being manifestly desirable and justifiable for reasons of economy and reliability.

B. Factors established by or under the control of government:

1. Each investor-owned company's or government entity's perception of its objectives and obligations as they are established corporately, or by laws governing franchises of private companies, or government-owned entities. These objectives and obligations include service and financial goals. Service goals for private and public utilities include meeting demand, safety and reliability of service, and reasonableness of cost of service. Financial goals for the private sector are rates of return on investment perceived necessary by each company for it to remain financially viable, considering current and long-term demand and supply costs. The general goal is the reasonable rate of return which each regulatory commission will approve. Financial goals for the public sector vary but generally include revenues necessary to remain financially viable considering current and long-term electric supply requirements. Financial goals for the public sector are sometimes influenced by increasing supply price to raise taxes for general government purposes.
2. Regulation of rates, or control of prices, which can be charged for electric supply. Delays in rate making proceedings and in approval of rate increases. Negative impact of low rates and delays on "coverage" and ability to finance new and replacement plant. (Regulation is about 90 percent at the state level and 10 percent at the federal level. Rate regulation is most pertinent to the regulated investor-owned companies.)
3. Regulation of design, construction and operation of generation and transmission facilities, at federal and state levels, and concerned with siting, environmental and health and safety matters. Regulation is

perceived in many instances as impairing ability to install new generation and transmission facilities required to meet current and future demand. Issues involved are requirements and administration together with associated delays.

4. Major uncertainties in selection, availability and price of fuels for use in generation, due in part to environmental regulations, and in part to general supply dislocations in the oil, gas and coal industries, some but not all caused by government law and regulation.

C. Factors established by or under the control of others:

1. Public attitudes, perceived to be troublesome, and interacting with federal and state executive, legislative and judicial processes and impairing the development and application of rational decision making by government in regulation of rates and environmental and nuclear safety matters.
2. Historically influential manufacturing firms such as General Electric and Westinghouse and their planning, research and development, manufacturing and total service capabilities.
3. Consulting and engineering and construction firms such as Bechtel, Stone & Webster and Ebasco and their capabilities for analysis and presentation of options, design and installation of new facilities, and modification of existing facilities.
4. All suppliers of goods and services, their competitive offerings, and their perceived capabilities and performance.
5. Major uncertainties in selection, availability and price of fuels (see B.4 above).

D. See also general conclusions on decision-making in the energy industry, Appendix A.

II. Observations on energy policy and the electric utility industry.

Policy, law and regulation should be developed and administered so as to provide for:

1. Economic factors.
   a. Returns on investment necessary to assure the service capability and financial integrity of the industry.
   b. Orderly and timely rate adjustments pursuant to applicable law and regulation so that the industry is assured of being able to act responsibly, considering finances and service.
2. Demand and price.
   a. Cost of service pricing to all categories of customers.
   b. Recognition of electricity as the prospectively dominant energy form of the future and consequent emphasis on structuring policy, law and regulation which permits its optimum economic production and utilization.

3. Restrictions on supply.
   a. Regulations governing health and safety, and environmental matters (including nuclear licensing and safety), and especially those influencing planning, design, construction and operating decisions, not being allowed to cause unreasonable delays in programs perceived by the industry as vital to its functioning in an orderly and efficient manner.
   b. Regulations governing the use of alternate fuels being based on a realistic view of the compatibility of the regulations with (1) the realities of supply and demand, and (2) the obligations of the industry to provide electric energy, especially regulations influencing price, assured availability and environmental acceptability.
4. Government as supplier.
   a. In these instances where government assumes the role of monopoly supplier, as it so far has in enriched uranium and technology related thereto, government acting as a reasonable, prudent, and fair supplier working in concert with and being responsive to the problems of its captive buyers.
   b. Government eschewing the role of monopoly supplier of any goods or services (electricity not referred to here) required by or available from the private energy industry except in the face of a market failure manifestly beyond the capability of any or all interested private and public organizations to correct.
5. Organization and industry structure.
   a. Reexamination of the structure of the industry, public and private, with the intent to rationalize the organization of both sectors as they relate to present and perceived national requirements for meeting demand, reliability and reasonable cost. Activation of the study called for under Section 30 of the Holding Company Act may suffice. If there is perceived merit to a smaller number of larger electric utility companies then the governmental disincentives to aggregation should be removed and the private sector should be encouraged to explore and carry out a program for aggregation into larger corporate entities; or demonstrate that the present organization mode is necessary and sufficient. Policy, law and regulation should allow, rather than discourage, mergers of private electric utility systems in those instances where sound management practice, economy and reliability would appear to favor larger systems.
6. Research.
   a. Recognition and utilization of the Electric Power Research Institute to plan and coordinate those national research and

development activities relevant to the electric utility industry which its private and public sponsors wish to support. Recognition that there is a continuing important and vital role in specific R&D activities for individual electric utility companies, electrical manufacturers, and others who may wish for their own reasons to undertake minor or major R&D programs independent of EPRI.

7. See also general observations on energy policy, Appendix B.

Appendix E

Influences on Decision-making in the Gas Industry

I. Decision-making in the gas industry is influenced principally by:

A. Factors established by or under the control of the industry:

1. Each company's perception of its objectives and obligations—principally return on investment.
2. Each company's perception of its objectives and obligations which may go beyond return on investment and/or gas supply service requirements.
3. In the instance of subsidiary companies, objectives and obligations are as dictated or approved by the parent company and the decision-making of the parent becomes dominant.
4. General recognition and demonstration that public utility service requirements cannot be met due to lack of supply—and acceptance of this as inevitable.
5. Each company's perception of present and future demand in markets available to it.
6. Each company's and industry group's perception of domestic natural gas and suitable alternate supplies of gas—LNG, syngas, imported (Canada), Alaska . . . available to it.
7. An industry-wide perception that near- and long-term demand vastly exceeds supply capability under any supply scenario, with little sensitivity to substantial increases in supply cost if "roll in" of new alternate supply is allowed by regulators.
8. An industry-wide perception that the much higher incremental prices of supply alternate to domestically produced natural gas must be "rolled in."

9. Each company's perception of pricing required to satisfy its objectives and to assure satisfactory response from the money market for additional capital required.
10. Each company's perception of competition from prospectively alternate energy supplies—oil, electric and coal.
11. Industry associations', such as the American Gas Association and the Independent Natural Gas Association, ability to develop and carry forth industry views and interact with the government.
12. Individual company and industry groups' research and development activities.
13. Lack of strong, resourceful, innovative suppliers of goods and services with interests closely and singularly related to those of the gas industry.
14. The diverse objectives and obligations of companies in the industry, generally but not always, separately divided into production, transmission and distribution companies with each company able to act in conformance with its own objectives and obligations. (The supply problems of the industry have been changing this, with distribution and/or transmission companies moving into production activities.)

B. Factors established by or under the control of government:
   1. In the regulated segments of the industry, the public utility service requirements dictated by franchises.
   2. Availability of research and development support.
   3. Relevant public policy, law and regulation and its administration, perceived by the industry to be the major factor in creating the supply/demand imbalance.
   4. Uncertainties on the quantity and price of supply deriving from the actions of the U.S. and state governments—especially those governing leasing and environmental matters.
   5. Uncertainties on the quantity and price of supply deriving from the actions of foreign sovereign nations—especially those governing availability and price of LNG and natural gas (Canada).

C. Factors established by or under the control of others:
   1. Public attitudes, perceived to be negative, interacting with the legislative and judicial process at the federal and the state level.

D. See also general conclusions on decision-making in the energy industry, Appendix A.

II. Observations on energy policy and the gas industry:

Public policy, law and regulation should be developed and administered so as to provide for:

1. Economic factors.
   a. Natural gas field price levels, regulated or not, which are deemed by the market to produce supply for sale at prices the ultimate consumers determine to be acceptable.
   b. Stability in sales contracts of producers committing new supplies to interstate markets.
2. Supply factors.
   a. Public land lease sales, especially those in offshore areas, in keeping with industry estimates of demand.
   b. Administration of environmental regulations relevant to the transportation and importation of natural gas and the production of synthetic gas which permits the orderly planning and execution of research and development, and supply programs.
3. Organization and industry structure.
   a. Encouraging mergers of distribution companies and further development of regulated integrated transmission and distribution with greater financial, manpower and technological capability to produce their own supplies and/or negotiate with major producers.
4. See general observations on energy policy, Appendix B, observations on energy policy and the oil industry, Appendix C, and on the coal industry, Appendix F. (These relate principally to supply of natural and synthetic gas.)

Appendix F

Influences on Decision-making in the Coal Industry

I. Decision-making in the coal industry is influenced principally by:

A. Factors established by or under the control of the industry:

1. Each company's perception of its own objectives and obligations—principally return on investment.
2. In the instance of subsidiary companies, objectives and obligations are as dictated or approved by the parent company and the decision-making of the parent becomes dominant.
3. Each company's perception of supply requirements—usually as specified under existing contracts or made the subject of specific contracts before new mines are opened.
4. The general perception of long-term major increase in demand for coal—for electric generation, gasification (low BTU gas for electric generation and high BTU gas for pipelines), and liquefaction to synthetic oil.
5. The perception that the future for coal is in the low sulfur, low cost, minimal health and safety and labor problems of the western strip mine coals; rather than in the high sulfur, high cost, difficult health and safety, and troublesome labor problems of deep-mine eastern coals. This is a major influence in seeking environmental regulations which would allow strip mining of western coals. (This perception is not shared by some newly active in the eastern coal industry, and others who own very large reserves which they expect to mine eventually.)
6. Competition within the industry, combined with ability to co-operate where there is no threat to the separable interests of the

cooperators—such as research and development and industry/government and industry/labor interactions.

7. The larger companies providing leadership, especially in industry/government (and perhaps in labor) interactions.
8. Industry associations', such as the National Coal Association, ability to develop and carry forth industry views and interact with government.
9. Individual company and industry groups' research and development activities.
10. General lack of strong financial and technological resources of the character required for significant innovative work on mining, including health and safety, and environmentally acceptable end use of coal. (This is true only at the coal company level, not in those instances where control is held by larger corporate entities with significant financial, managerial and technological resources. To what extent these have yet been brought to bear on coal company operations is not clear to the author. Until the perceived return on investment is adequate, those resources probably cannot be applied to coal.)
11. Lack of strong, resourceful, innovative suppliers of goods and services with interests closely related to those of the coal industry.
12. The capacity and cost of rail transportation, these perceived as being unsatisfactory due to the general decline of the railroad industry.

B. Factors established by or under the control of government:
   1. Availability of research and development support.
   2. Federal and state environmental regulations governing mining of coal—especially strip mining; the use of coal—especially SO_2 emissions from utility boilers. Major uncertainties resulting from existing and pending legislation and influencing the acceptability of coal now being produced for electric utilities and the ability to mine alternate coals.
   3. Federal and state mine health and safety regulations of recent vintage; difficulties in administration by government, and conformance by industry.

C. Factors established by or under the control of others:
   1. Competition with alternate fuels (uranium, gas, oil) especially for electric generation use.
   2. Well-recognized and continuing public antipathy to the coal industry due especially to a perception of past misuse of both labor and the environment.
   3. Labor relations—historically poor and still so.

D. See general conclusions on decision making in the energy industry, Appendix A.

II. Observations on energy policy and the coal industry:

Policy, law and regulations should be developed and administered so as to provide for:

1. Economic factors.
   a. Recognition of coal as the U.S. principal long-term fossil fuel resource base.
   b. Revenues perceived by the industry and the money market to be adequate, with reasonableness of rate of return on investment being implicit.
   c. Price and tax policies applied to mining which are compatible with the current need for coal, and accelerated development and use of coal as the principal long-term fossil fuel resource base.
   d. Price and tax incentives applied to the end uses or products of coal (gas, liquids) which are compatible with the accelerated development and availability of coal derivatives as substitutes for other fossil fuel supplies—such as natural gas and oil.
2. Supply and use factors.
   a. Environmental restrictions on use of high sulfur coal being consonant with rational economic alternate fuel supply options, and/or proven technology.
   b. Mine health and safety regulations being reanalyzed with the intent to produce the necessary results on health and safety. (One suggested approach is: (i) general standards be set by government in consultation with industry, (ii) industry be assigned the full responsibility and authority for assuring mine health and safety with serious penalties for non-conformance, and (iii) Bureau of Mines' role in detailed supervision of industry mining operations be modified.)
   c. Environmental regulations governing mining of new surface coals based on federal guidelines being developed in consultation with industry, with each state developing and monitoring its own standards consistent with federal guidelines. Making conformance with state standards a prerequisite for permit to operate in a state.
   d. Development and production activities in coal mining, gasification, stack gas cleanup and other programs for the safe, environmentally acceptable and efficient mining and use of coal by providing, as necessary, special ad hoc price or tax incentives which result in a return on investment perceived sufficient by individual companies with relevant interests and capabilities; and administering leasing policies and environmental regulations to permit rational industrial programs.

3. Development of synthetic fuels.
   a. Development of a U.S. synthetic hydrocarbon fuels supply base as necessary to provide options for price leverage on, and ability to be independent of, foreign oil sources.
   b. Development and production of oil from shale and/or coal by providing, as necessary, special ad hoc price or tax incentives which result in a return on investment perceived sufficient by individual companies with relevant interests and capabilities; and administering leasing policies and environmental regulations to permit rational industrial programs.
   c. A program to assist in providing water requisite for synthetic fuels development and production programs, consistent with a balanced allocation of water to other important needs.
4. Industry organization and structure.
   a. Encouraging all companies with relevant interests, resources and capabilities not now active in the coal industry to engage in coal mining and development of uses of coal compatible with national energy supply goals.
   b. Encouraging the agglomeration of coal-oriented companies which have the requisite financial, managerial, technological and manpower resources to develop competently and acceptably the prospective greatly increased production and use of coal.
   c. Removing restraints to restructuring the industry into fewer, larger, more viable corporate entities.
5. Labor.
   a. Bringing management/labor relations more in line with U.S. norms. (For example, encouraging coal companies to assume directly the financial costs of major benefits now paid through the United Mine Workers.)
6. See also general observations on energy policy, Appendix B.

Appendix G

Corporate Objectives

As part of this study, definitions of corporate objectives were sought from leading companies. A major international oil company supplied a statement approved by its Board of Directors in 1972 for the period 1973-77. The emphasis of the objectives list, which is preceded by a series of background assumptions, is thought to be generally representative of the oil industry. The full statement follows:

BACKGROUND ASSUMPTIONS

As an introduction to our summary of this year's Objectives, we will first highlight the assumptions, both of a general nature and within the oil industry, that have influenced our approach. I will also note areas of uncertainty facing the oil industry and for which specific assumptions have not been made because of lack of definition.

A key change this year compared with prior years is the new economic policy put into effect by the U.S. in August last year which represented a sharp reversal with the past. Domestically, the U.S. imposed the first peacetime controls over wages and prices to reduce inflation, and at the same time adopted strong expansionary measures to stimulate the economy. Internationally, the U.S. abandoned the temporizing approach to seek a more fundamental solution of its balance of payments problem.

The effect of these policies will be a boost in short-term economic growth in the U.S. and to dampen growth overseas. However, long-run economic growth will be only marginally affected and with the exception of Japan will remain about the same as in prior year's assumptions. In the U.S., petroleum demand is forecast to grow at 4.5 percent per year compared with 3.4 percent forecast last year. In Europe, petroleum demand is forecast to grow at 7.5 percent per year compared with 8 percent last year, while in Japan petroleum demand is forecast to grow at 10.5 percent annually compared with 12 percent last year.

Inflation both in the U.S. and overseas will be reduced from the very rapid rates experienced recently. The U.S. is ahead of the other industrial nations in achieving improvement in its current rate of inflation. We forecast that the rate of inflation in the U.S. will decline from 4.5 percent in 1971 to somewhat less than 4 percent in 1972. For the 1972/1977 period we forecast inflation to average 3.7 percent per year. Inflation in Europe which is currently continuing at very high rates of 5 to 8 percent per year is expected to moderate gradually over the next few years.

The currency values resulting from last year's realignment may not fully reflect the inflationary trends and international competitive situation as they develop or even as they exist in each major country. In particular, the current value of the Japanese yen may still be below the competitive level with the U.S. dollar, and the Japanese yen may have to be revalued. Occasional speculative pressure on the dollar may also occur in the near-term in Western Europe. However, in the long-term, inflation rates in Europe will generally be as high or higher than in the U.S. and confidence in the dollar can be expected to strengthen relative to European currencies.

We do not expect any new significant tariff cutting initiatives comparable to the Kennedy Round. As a result, during this period, growth in world trade is likely to be slower than during the 1960s and is not likely to increase its share of world gross national product.

Another key factor affecting the outlook period is the growing awareness of a shortage of natural gas and crude oil in the U.S.

Imported oil will provide nearly the entire growth in U.S. petroleum consumption through 1985. Synthetic oil production in the U.S. probably will not be significant until 1985.

Similarly, we project an increasing gap between conventional U.S. natural gas supplies and the potential demand for this fuel. Introduction of North Slope gas probably will not be sufficient to offset the decline in U.S. production.

A further key assumption to our Objectives development is that there will be no major change in U.S. tax policy. Accordingly, the Objectives reflect a continuation of the investment tax credit and no new measures such as contained in the Burke/Hartke proposed legislation which would disallow foreign tax credits on foreign earnings.

The Objectives contain no specific provision for participation by producing governments in existing concessions. Also, the Objectives contain no specific provision for U.S. flag requirements on vessels in U.S. import trade although legislation is pending which would require 50 percent of such trade to be in ships built in the U.S. and manned by U.S. crews.

CORPORATE OBJECTIVES

Our review of Corporate Objectives this year leads us to recommend one minor change. In focusing new investments where we can, as a minimum, by fully competitive, we have added the reminder that we must be competitive in key functions, as well as geographic, product and class of trade areas. This is designed to help emphasize the fact that in some geographic areas we may be at a competitive disadvantage in some functions, but not in others.

This year we also reviewed the Corporate Objectives as they relate to the longer term beyond 1977. As a result of long-range studies we can clearly see the conventional oil and gas business in North America leveling out and beginning to decline in the late 1970s with the growth of energy being supplied by imported oil, nuclear power, coal and synthetic production. The same pattern will obtain in the foreign area beginning at least by 1990-2000.

We concluded that our primary objective of a *X*[a] percent annual earnings growth is probably unattainable over the long-term without substantial participation in the synthetics industry and perhaps in other diversification activities. This conclusion, of course, is merely an extension of the assessment which has already led us to take smaller steps into uranium, coal and oil shale. Acquisition of sizable coal reserves in the U.S. should be a prime objective. In the longer term, we should also examine the possibility of acquiring suitable coal reserves overseas (for example Australia).

Certain of the key supporting Corporate Objectives will also be affected in the long term.

- Improvement in rate of return on assets will be difficult to achieve if we extend ourselves into uranium and synthetic fuels.
- Opportunities to increase earnings and assets in North America should be available, since the U.S. has an abundant resource base for synthetic fuel.

CORPORATE OBJECTIVES

The corporation has the objective to maintain an average annual growth in earnings per share of at least *X*[b] percent. The following are the major supporting objectives:

1. Improve the rate of return on assets for the Corporation to the average of our major competitors.
2. Continue to search for opportunities to increase earnings associated with producing operations in North America, while at the same time ensuring access to economic reserves of crude oil and natural gas.
3. Continue to search for opportunities to increase earnings and assets in North America.
4. Achieve material diversification of Corporate earnings.

---

[a] Precise figure, which fell between 5 and 10 percent, deleted at request of the company.

[b] Precise figure, which fell between 5 and 10 percent, deleted at request of the company.

5. Eliminate existing assets that do not and cannot provide an adequate rate of return to the Corporation.
6. Focus new investments in selected functional, geographic, product and class of trade areas where the corporation can as a minimum be fully competitive.
7. Maintain secure access to our customers so that the Corporation will be in a position to maximize its long-range profitability in the face of severe competition.
8. Increase the returns on our investments in capital equipment and trained manpower by an unremitting drive to maintain tight control of costs and expenses and improve operational efficiency.
9. Maintain access at all times to new developments in technology; have unique knowledge in a selected number of technological areas that offer prospects for above-average returns; and translate these developments expeditiously into profit opportunities for the corporation.
10. Ensure that the implementation of the Corporation's objectives reflects consideration of the need to conserve the physical environment.

Appendix H

Electric Utility Financing

INTRODUCTION

The failure of Consolidated Edison to pay its common stock dividend in the spring of 1974 brought to widespread public attention a problem that had been growing in the utility industry for several years—inability to attract sufficient capital to permit the expansion necessary to meet increasing demand for electricity.

The following is excerpted from a position paper by Eugene W. Meyer, a Vice President of Kidder Peabody & Co., Inc. which was circulated within the public utilities industry in 1974.

. . . The deteriorating utility financial situation has existed since the late 1960s, but the Con Edison dividend omission has caused investors to focus on the industry picture and turn away from further investment.

Of the 4,415,315 new shares of common stock scheduled to be offered by the electric utility companies in the month following Con Ed's April 23 bombshell, only 765,315 have been sold as of this writing. Thus, only $13.8 million out of $98.5 million required has been raised. Furthermore, another 20,140,000 shares totaling $356 million is scheduled for the remainder of May and June; 15,600,000 shares totaling $216 million is scheduled for July; 21,100,000 shares totaling $345 million is scheduled for September and another $1 billion of new equity will be required during the fourth quarter. Few of the companies scheduling these offerings are able to substitute debt or preferred stock offerings in lieu of common stock without suffering further bond rating reductions and/or without violating outstanding bond indentures and corporate charter restrictions.

What this says is that a large segment of the electric utility industry (basically those companies serving the heaviest population concentrations) are now at the crossroads of inability to finance and immediate action is required to attract private equity investment capital back to this industry.

Origins of the Problem

A combination of factors has been working to create the present extremely serious problem for utilities, particularly electric utilities. First, costs have been rising at a rapid and increasing rate. Second, due to fuel shortages, conservation measures have been introduced which have reduced revenues. Third, regulatory commissions have been unable or, in some cases, unwilling to adjust rates rapidly enough to compensate for this combination of rising costs and reducing revenues. Fourth, despite their inadequacy from the point of view of the financial viability of utilities, rate increases have been substantial and have been meeting increasing public and political resistance. Fifth, since utility capital structures are highly leveraged on the premise of steady earnings due to regulation, the adverse effect of these factors on earnings has been magnified. Sixth, the present problem is further exacerbated by the fact that it comes after several years of more gradual deterioration in the utility financial picture due to (a) enormous construction expenditures required to meet demand, (b) higher costs of capital, (c) inflation and (d) rate increases that failed to keep pace with increases in capital and operating costs. Finally, this crisis comes at a time when the market for common stocks of all companies is in a very bad slump as set forth above.

We will describe in some detail the present problem in the market for common stocks of utilities (Part II) as well as the origins of the present problem (Part III) and will conclude with some proposed solutions (Part IV).

II. THE PRESENT PROBLEM

Construction Engendered Capital Needs

Since at least 1969, the utility industry has been faced with sharply rising construction expenditures. The investor-owned electric utility segment of the industry has gone from $8.3 billion of construction expenditures in 1969 to $14.9 billion in 1973. Current estimates are for construction expenditures of about $16 billion in each of the years 1974, 1975 and 1976, increasing to $18 billion in 1977 and $19 billion in 1978.

At the same time, internally generated funds from retained earnings, depreciation, and deferred income tax increased only from $3.1 billion in 1969 to an estimated $5.3 billion in 1973.

Capital Markets

Since 1969, the rising external capital requirements of the electric industry have caused a tremendous increase in the amount of new common stock which was required to support the even greater amounts of long term debt and preferred stock. New common stock must be sold to maintain the debt-equity ratios required in the industry. In 1969 the industry issued $611 million common stock; in 1970, $1.6 billion; in 1971, $2.4 billion; in 1972

$2.5 billion and in 1973, $2.6 billion. Recognizing the tremendous dilution implied by these new equity requirements, institutional investors reduced their utility stock holdings substantially and, since last year, there has been virtually no institutional interest in utility common stocks. Since 1969, therefore, the vast majority of new common stock has been sold to individuals who have been primarily interested in the current yield provided by the common stock dividend. Those investors continued to hold utility common stock and purchased additional shares on the assumption that there would be no dividend cuts in the utility industry. They maintained this belief, in spite of the steadily deteriorating financial picture of the industry, because they had faith that the regulatory system would not permit the financial soundness of the industry to be jeopardized.

The utility industry's capital-raising dilemma has been aggravated by the sharp advance in interest rates. The increased amount of debt and preferred stock financing required in the industry's construction budget at these higher interest rates resulted in an increase in the combined interest and preferred dividend cost as a percentage of total operating revenues from 10.5 percent in 1969 to 13.2 percent in 1973. At the same time, fuel costs, as a percentage of total operating revenues, increased from 17.3 percent in 1969 to 24 percent in 1973. The regulatory system has been unable to react quickly enough to increase the rates to offset these rising costs. As a result, the rate of return on common equity for the industry has declined from 12.73 percent in 1967 to 12.16 percent in 1969 to 11.46 percent in 1973. Consistent with that performance, utility common stock prices have also declined steadily.

Con Ed Dividend Omission and Aftermath

For about two months prior to Consolidated Edison's decision to omit the dividend, securities salesmen were getting an increasing number of inquiries as to the safety of the dividends of utility companies. Suspicion in this area was making it ever more difficult to market new issues of common stock for the industry. When Consolidated Edison announced its dividend action, it served to confirm the worst fears that had been growing in the individual investor's mind. As a result, there has been heavy selling in utility common stocks since the Consolidated Edison action. The effect has been to drive the bulk of the common stock prices of companies in the industry to levels substantially below book value at a time when the industry is faced with a need for even larger amounts of new equity. As of the close of the market May 14, 1974, forty-four of the fifty-one largest electric utilities in the nation were selling at prices below book value. The average of the fifty-one companies was a market price 78.8 percent of book value.

Even at these new lower prices, there has not been any rekindling of interest on the part of the institutional investors. The institutional investor recognized that, as utilities issue ever larger amounts of common stock below book value, thereby diluting both earnings per share and assets per share, the financial plight of the utilities will worsen.

What dilution means to existing stockholders is illustrated in the following example. Assume a company with 1,000 shares outstanding and a book value per share is $10. Assume the market price of the stock is only $7 per share. If the company now sells 100 shares of stock at a price of $7, the total book equity of the company will rise from $10,000 to $10,700. But since there are now 1,100 shares outstanding, the book value per share has declined from $10.00 to $9.73. Each of the previous investors has lost 3 percent of his investment. If the company is a public utility whose earnings are proportioned to its book equity, the investor has had a 3 percent decline in earnings per share as a direct result of the utility's sale of the stock, even assuming that the rate of return on equity is not at all eroded.

Utility financing has been based on the fundamental assumption that long term financing will always be forthcoming, thereby enabling the companies to finance short-term requirements with large amounts of short-term debt and to operate with the least cash possible. This method has been endorsed by the regulatory commissions because it meant lower rates to the consumer. However, it is axiomatic that when a utility company is unable to sell new common stock, it is only a short time before it is also unable to sell long term debt securities because there is insufficient equity to back up the debt. Since the Consolidated Edison decision, Toledo Edison was forced to reduce the size of its new offering of common stock from 800,000 shares to 400,000 shares because a market did not exist for any more than 400,000 shares. The next company scheduled to offer common stock was Consumers Power Company, with an offering of 2.5 million shares. This offering was cut to 1.5 million shares when it became apparent there was not a market for 2.5 million shares. When the actual offering date came, however, Consumers Power announced that it had postponed its offering due to a problem at the Palisades Nuclear Plant. However, it is common knowledge in the investment community that there was no market for 1.5 million common shares on any kind of reasonable terms.

Consolidated Edison itself has, of course, withdrawn its scheduled $150 million equity offering. It is doubtful whether Consolidated Edison could sell new common stock at any price. Long Island Lighting Company, feeling the impact of the current market situation, cancelled a common stock rights offering scheduled for the last week in May.

The Present Outlook

It is clear that additional credit cannot be extended to this industry if it continues its construction program and is not able to attract equity capital. Furthermore, massive cutbacks in construction programs would result in substantially increased unemployment in the construction area of the economy and also in electrical equipment manufacturing. Such construction cutbacks would also adversely affect the reliability of the nation's electrical system for years to come. Adequate capital to finance the needed construction programs for this industry is present in our economy, but the industry in its current financial posture is to a greater extent than is generally realized unable to attract the needed equity investment. The equity capital that can be sold will have a very high cost which ultimately must be borne by the ratepayer.

If this industry is able to improve the cash flow quality of its earnings to the point where investors will once again feel that the common stock dividends are indeed secure, then utility common stock prices will increase to levels at or around book value and the industry's future equity requirements can be met. However, regaining the confidence of the equity investor will require a substantial increase in the industry's rate of return on equity, which can only come about by seeing to it, at the regulatory level, that the prices at which electricity is sold in this nation cover the costs, including the costs of capital, which are incurred in providing electric service. Other measures to attract individuals and institutions back into the market for utility common stocks are discussed in Part IV.

III THE ORIGINS OF THE PRESENT PROBLEM

Leverage

Utility capital structures are characterized by a high degree of financial leverage. The use of financial leverage by utilities is in turn premised on the principle that through rate adjustments utility earnings available to common stock will be steady and subject to less fluctuation than earnings of other companies. However, when rates are not adjusted rapidly enough to cover fast rising costs or where revenues are reduced because of reduced consumption, the downward effect on earnings available for common stock is magnified because of the high degree of leverage. In recent months, utilities have been experiencing both rapidly rising costs, especially of fuel, and reduced consumption necessitated by the Arab oil embargo.

The effect of reduced consumption and rising costs is further magnified by the high degree of operating leverage characteristic of utilities. That is, utilities have relatively high fixed costs and lower variable costs—it is a highly capital intensive industry. The high degree of operating leverage results primarily from operating and not financial considerations, but the effect in times of reduced consumption is no less marked.

Construction Requirements

Because of their franchise from the State, utilities, unlike other corporations, have a legal obligation to provide service to meet the demand for electricity. The growth in demand has required an enormous increase in generating capacity. Even though conservation measures have reduced current demand in general, the present forecasts indicate no appreciable drop in peak demand on which construction programs are necessarily based.

Over the past five years, the annual rate of growth of electric sales has been about 7 percent. While declines in the rate of population growth, environmental concerns and the rising cost of electricity are gradually reducing the rate of growth, the impact of these factors on the industry's construction program in the next few years will be minimal. Primarily because of siting problems, the lead time on plant construction is now about eight years (longer for nuclear) so that the construction program into the early 1980s is largely committed.

The combination of growth and rising costs means that the cost of new utility plants 1973-82 will probably exceed $200 billion (electric utilities require about $4 of plant to produce $1 of revenue versus $.25 of plant for each $1 of revenue by the automobile industry.) Since, as noted, less than half of the necessary funds can be generated internally, the electric utility industry will have to raise more than half of the $200 billion in the market. On top of that amount, about $14 billion of bonds will be reaching maturity during the balance of the 1970s and must be refunded.

Adjusting Rates to Cover Costs and Required Return on Capital

In the years prior to the late 1960s, ever growing economies of scale in electric generation more than offset cost inflation so that the cost of electric service showed a steady decline. However, beginning in the early 1970s cost inflation began to outstrip economies of scale and the cost of electricity began rising steadily and substantially.

At the same time that inflation began to have a serious adverse impact, the cost of capital also rose dramatically. Interest rates on rated electric and gas utility long term bonds went from about 4.5 percent in 1965 to as high as 9.5 percent and they are currently being marketed at about 9 percent. Because of the enormous amount of debt securities issued in this period to finance rapid expansion, the imbedded cost of debt went from about 4 percent in 1965 to about 7 percent in 1973. Dividend rates on new issues of preferred stocks have had a similar rise to over 9 percent currently.

While the cost of common stock is essentially related to a combination of dividend yield and growth rate, utility (gas and electric) stocks traditionally sell on a yield basis. Yields on common stocks have gone from 3.6 percent in 1965 to a current level of over 8 percent. Wages and other costs other than fuel have increased substantially. The quantum jump in fuel costs is too well known to require extensive discussion.

Apart from inflation, other factors, primarily environmental, have added about 3 percent annually to the cost of electricity. Moreover, the economies of scale, especially in terms of thermal efficiency, are tending to peak out. There is relatively little room for improved efficiency in current fossil fuel generating technology and it is apparent that improved nuclear fuel efficiency will be attained only at greatly added cost.

Effect of Energy Conservation on Construction

There is little indication that energy conservation measures will greatly reduce present or future construction budgets. Present construction is basically committed and there are not economic or operating reasons to delay placing plants currently under construction on line. Much planned construction is needed for fuel switching, reliability and to make up for present deficits in supply (as evidenced by the voltage reductions of 5-8 percent in New York City in the summer of 1973). In addition, there is no assurance that energy conservation efforts will affect system peaks and indeed such peaks already reflect voltage reductions for conservation reasons.

Ratemaking Developments

Public utility rate regulation is comprehensive and essentially cost related. Unless rising costs are promptly covered by rate increases so as to provide a steady flow of earnings, the fundamental assumption on which the capital structure of utilities is based is lacking. The two critical problems of rate regulation today are (1) the time interval between cost increases and rate increases and (2) a fair rate of return to the equity capital.

In the first area, regulatory lag, there has been by traditional bureaucratic evolutionary standards great progress. Most state regulatory commissions have accepted the concept of interim rate increases (too many, however, restrict their use to "emergency" situations), some (but nowhere near enough) permit projection of cost increases, and automatic rate adjustments for fuel costs and, in some cases, other costs, such as taxes. However, the accelerated rate of increase of all costs has still left the rate schedules far behind the current costs of providing energy.

While regulatory lag exacerbates the problem of a fair return on equity, even if there were no regulatory lag, that problem would require serious attention. Based solely on rate of return on equity allowed in rate cases (putting aside the fact that even that rate is seldom if ever earned because of rapidly rising costs), rates of return on equity have lagged behind rates for debt and preferred (which are set by the market).

During recent years (1965 to the present), utility common stock prices have declined substantially more than general market averages; the ratio of market price to book value of electric utilities common stocks has declined drastically; and electric utilities' earnings coverages have fallen drastically (in some cases to the point of precluding the issuance of additional senior securities). Although all of these factors call for substantial increases in the allowable return on common equity, many regulatory commissions are only allowing a return on equity of as low as 11-12 percent.

IV. SOME PROPOSED SOLUTIONS

The proposed solutions fall into two categories, ones that can be effected at the federal level and those that must, unless there is a total preemption of the state regulatory function (a step which is neither practical nor desirable), be carried out by the states individually.

On the federal level, the proposals take the form of relatively slight adjustments in certain tax provisions to somewhat increase utility cash flow.

On the state level, affirmative steps should be taken to insure that adequate rate relief is granted to offset the combined effects of rapidly increasing costs and conservation-induced revenue reductions applied in an industry that is highly leveraged both from a financial and operating point of view.

Federal Tax Changes Proposed

The three changes in the federal tax law proposed to increase utility cash flow are:

1. Eliminate the disparity introduced by the 1969 Tax Reform Act between utilities and other companies by increasing the investment tax credit from 4 to 7 percent.
2. Reinstate the provision in the federal tax law prior to 1969 that permitted the use of accelerated depreciation in calculating earnings and profits for the purpose of determining the proportion of dividends constituting a nontaxable return of capital.
3. Extend for five years the availability of rapid amortization of pollution control equipment.

Serious consideration should be given to conditioning the investment tax credit increase on state regulatory action to normalize the accounting for such credit in the same way the use of accelerated depreciation was conditioned in the 1969 Tax Reform Act.

State Regulatory Changes

What is needed in the state regulatory area (and which the Treasury can support in policy statements, meetings with state officials, etc.) is uniform state laws authorizing the use of projections, temporary rates, statutes with limited periods in which the agency must act, and further use of automatic adjustment clauses.

As a longer term solution, we urgently propose that these important changes occur in each regulatory jurisdiction:

1. Eliminate the use of "flow-through" accounting for ratemaking purposes.
2. The industry's annual depreciation rate must be increased at least a full percentage point.
3. The accounting method for "allowance for funds used during construction" must be gradually eliminated, substituting instead construction work-in-progress in the rate base for ratemaking purposes.

These three proposals would substantially increase cash flow along the lines of other industries. None of the proposals would increase earnings to the common stockholder but the increased cash flow would increase the confidence that utility common stock dividends are, indeed, secure. All of the proposals would, however, require rate increases over and above those necessitated by rising fuel costs and rising interest rates. It is even possible that the net effect of these three actions would result in increased federal income taxes for some companies. Because of this and because of the rate increases required, it is suggested that the state regulatory bodies be given three years to implement these changes by including the changes in normal rate cases as they occur. The federal government should, however, monitor the states to be sure of progress. The federal government should also assure the states that they do not encourage these measures merely to increase federal income tax revenues. Therefore, utility company effective federal income tax rates should not be allowed to exceed December 31, 1973 levels for a period of six years.

The above proposals would lead to rate increases to be sure,but not to the full amount involved. The changes would cut external financing requirements nearly in half. Thus, far less new money would be raised at prices well above imbedded costs, resulting in a lower cost of capital.

Finally, these proposals would reduce somewhat federal tax revenues. That is a serious problem but the alternative is for government takeover of the industry which would require tremendous increases in taxation and government borrowing. Furthermore, it should not be forgotten that virtually everyone pays both electric rates and taxes.

Appendix I

Resource Needs for Nuclear Power

Increasing U.S. power demands coupled with declining supply and increased costs for conventional fuels will place steadily increasing demands on the nuclear power industry during the latter part of the 20th century.

As the industry seeks to meet these demands it faces a broad spectrum of problems—technical, economic and political.

In early 1973, W. Kenneth Davis, Chairman of the Atomic Industrial Forum's ad hoc Committee on Resource Needs for Nuclear Power Growth, Summarized the major obstacles facing the industry in a presentation to the Congressional Joint Committee on Atomic Energy. Following are excerpts from his presentation:

In view of the increasing concern about the supply-demand relationship for energy in the United States and the critical problems posed by the projected necessity to import 10 to 15 million barrels per day of oil into the U.S. in the mid-eighties (largely from the OPEC [Organization of Petroleum Exporting Countries] countries), the Executive Committee and Board of Directors of the Atomic Industrial Forum established in September 1972 an ad hoc Committee on Resource Needs for Nuclear Power Growth to investigate the potential contributions that nuclear power might make to alleviate the situation. Further, the Committee was asked to identify and make recommendations concerning problems that might appear to be obstacles to this maximum feasible utilization of nuclear power.

In considering the possibility of accelerating the nuclear power program, the Committee decided to look at two basic growth curves as "tests" of the problems involved. The first, or "reference case," is the projection which until recently has been generally accepted and widely used by the AEC and industry: 145 GW [gigawatts] by the end of 1980, and 300 GW by the end of 1985, and 500 GW by the end of 1990.

The second case, called the "expanded case," was selected as somewhere near a maximum within the constraints of the overall growth of electric power. While the same as the "reference case" in 1980, the "expanded case" totals 365 GW in 1985–an increase of 65 GW, and 700 GW in 1990–an increase of 200 GW. It was felt that the period between now and the end of 1980 was too short to permit any additions. Indeed, current estimates by the AEC and others now project a *decrease* of about 25 GW in the nuclear capacity to be installed by the end of 1980. We would hope that some way may be found to "recoup" that 25 GW slippage.

Based on these two cases, order of magnitude quantities were estimated for uranium requirements, uranium enrichment capacity, plant components, sites, fuel handling and processing rates, etc., to serve as a conceptual framework for assessing the problems that might be created or aggravated by the incremental additions.

Areas considered within this conceptual framework included; power plant siting; utility system planning, operations, and power distribution; licensing and environmental impacts; raw materials (uranium, thorium, zirconium, etc.); uranium enrichment facilities; nuclear fuel manufacturing; NSSS [Nuclear Steam Supply Systems] systems, equipment and components; engineering and construction; turbine-generators and power plant equipment; fuel transport and reprocessing; waste transport, storage and disposal; plutonium recycle; reactor safety; insurance and indemnity; antitrust; financing and power sales revenues, foreign investment and controls, etc. In carrying out the study, consideration was given to a mix of light-water and gas-cooled reactors.

Instead of attempting to study each of these areas in detail, we elected to identify those which might either be a fundamental limitation on the growth of nuclear power or which might prove to be "bottlenecks" if action were not taken in the immediate or near-term future. Those which could be handled in the normal course of events as, for example, the installation of additional fuel reprocessing plants, were only considered to the extent that they might contribute to overall financing requirements, demands for engineers and craft construction workers, etc. The critical areas were then examined in more detail to ascertain the nature and magnitude of the problems and to try to develop recommendations for avoiding or surmounting them.

At the outset, we made several assumptions which would serve as conditions in the conduct of the study. It was decided that priority attention would not be focused on the desirability or justification for an expanded nuclear power program–that attention would be focused on the feasibility of such a program and on the consequences of implementing it. We assumed that the national demand for energy would continue to grow and that the "expanded case," simply as a test case, would fit within the general limits of what might be met by electric power. We assumed that the breeder reactor, while possibly being brought to commercial use during the time period encompassed by the study, would not have a large enough impact during that time period to significantly influence the conclusions reached.

One of our basic assumptions was that problems plaguing the licensing process would be solved to the extent that the planning, design and construction of power reactors could be carried out on a rational, orderly, predictable and reasonable schedule. However, it was perhaps inescapable that the "assumption" would become one of the principal "if's" in the whole endeavor.

In order to get some feeling for the extent to which acceleration (or even achievement) of the nuclear power program would help alleviate the energy supply problems in the United States, it is useful to examine the equivalent quantities of fossil fuel which would be saved by the nuclear program. Based on annual use factors of 80 percent and equivalent full-power lifetimes of 17½ years, the following figures can be estimated;

1. The 25 GW "shortfall" in the nuclear program by the end of 1980 is equivalent to:
   770,000 barrels of oil per day
   6¼ billion barrels of oil (lifetime), or over half of the estimated oil reserves in the North Slope of Alaska (10 billion barrels).
2. The 365 GW potential nuclear capacity by the end of 1985 is equivalent to:
   11 million barrels of oil per day, 90 billion barrels of oil (lifetime), or over twice the present power reserves in the U.S., including Alaska (30 billion barrels).
3. The "potential acceleration increments" of 65 GW by the end of 1985 and 200 GW by the end of 1990 are equivalent to:
   2 and 6¼ million barrels of oil per day.
   16 and 40 billion barrels of oil (lifetime).

While the 2 million barrels per day "gain" in 1985 may not be large compared to the estimated imports of 10 15 million barrels per day in the 1980s, it is nevertheless very substantial and could play a significant part in closing the "gap." Without nuclear power at all, the "gap" could be doubled!

CONCLUSIONS

By far the most important conclusion of the study is that it appears to be *feasible* to build and operate nuclear power plants to accommodate any desired level of production up to the "expanded" case studied and within the general constraint of overall electric power growth–in the period after 1980.

However, this feasibility and the schedule on which nuclear power plant installation might be accomplished depends critically on several problem areas requiring early and decisive action. Unless these problems are resolved, it will not be possible to achieve the "expanded case" type of results, or unfortunately, even the degree of nuclear contribution assumed by the "reference case."

The principal "bottlenecks" which I will identify and discuss are in the following areas:

1. licensing;
2. uranium supply;
3. uranium enrichment;
4. technical and construction manpower;
5. financing.

LICENSING

Licensing requirements for nuclear power plants have changed dramatically over the past three years. Prior to 1970, licensing was focused on two related issues; reactor safety, and control of radioactive emissions to the atmosphere and to receiving bodies of water. Recent legislation and court decisions now require two additional issues to be considered; environmental impacts and antitrust implications.

Further, ill-defined and uncoordinated mechanisms evolved from a continuum of changes to a process set up nearly 20 years ago under a totally different set of circumstances, no longer lend themselves to the expeditious licensing of nuclear power reactors. Indeed, it is the delays in the licensing process that have put in jeopardy the 25 GW by which we may miss attainment of the long-held projections of 145 GW by 1980. If improvements in the process are not shortly forthcoming, there is serious doubt that the nuclear industry will be able to meet the 1985 and 1980 projection goals of the "reference case," much less those of the "expanded case." Some people in the industry believe that short of a major overhaul of the process, neither projection will be met.

It should be possible to reduce the lead time for nuclear units; i.e., from the date of utility-owner commitment to commercial full-power operation, to seven years. This is in contrast to the 9-10 years now dictated by licensing uncertainties.

A seven-year schedule has been accomplished in the past in the U.S. and is in line with current practice in other parts of the world, including Europe and Japan. A lead time of seven years should permit;

- Selection of a site, including meaningful public participation.
- Negotiation and implementation of commercial and contractual arrangements.
- Regulatory review of reactor safety, environmental impacts and antitrust implications.
- Procurement of equipment and on-site construction.
- Pre-operational testing.

At a very minimum, a streamlining of the current licensing process at both the administrative and legislative levels is called for, including:

- Amending the Water Act to make clear the: (1) federal agencies may commence formal review of site-related applications prior to state certification; and (2) a discharge permit that will cover the first five years of operation may be issued in advance of construction.
- Encouraging agencies with site-related responsibilities to coordinate their reviews and to conduct joint hearings.
- Developing legislation to assure a coordinated review of site-related issues and to provide a "one-stop" review and approval mechanism.
- Adopting an early and coordinated site review and approval step which would settle—in advance of AEC review of reactor safety features of the plant—the acceptability of a site, and establish any necessary environmental restrictions on its use. Such a step would in time remove much of the licensing process from the critical path of individual projects inasmuch as it would have established an inventory of sites whose availability was known in advance.
- Clarifying the implementation of NEPA [National Environmental Policy Acts] by vesting responsibility in a "lead agency" and committing to CEQ [Council on Environmental Quality], rather than to the courts, the responsibility for issuing interpretive rulings on the scope and conduct of NEPA reviews.
- Permitting AEC to issue a construction permit, subject to subsequent resolution of pending antitrust issues.
- Making more efficient use of agency review staff by eliminating duplicating reviews and coordinating reviews of common issues. A related step that should be taken is to relieve the ACRS [Advisory Committee on Reactor Safeguards] of mandatory review of each and every individual application.
- Seeking innovative mechanisms to reverse the current trend toward over-judicialization of the administrative processes employed in reactor licensing. We endorse the AEC's recent use of rule-making proceedings for the resolution of generic issues.

URANIUM SUPPLY

Although there is no reason to believe that sufficient resources of uranium cannot be found to meet the requirements of an expanding nuclear power economy, the Committee has serious reservations that unless ways and means are found for greatly expanding the accelerating exploration efforts, sufficient reserves will not be found in time to support the "reference" or "expanded" case projections.

The fact that an eight year forward reserve of domestic uranium adequate to meet the needs of the nuclear power industry will exist at the end of 1973, even if no added reserves are found in 1973, is dangerously misleading. Exploration expenditures in the U.S. in 1972 were less than they were in 1971, and it is likely that they will be even less in 1973. Discoveries in 1972 of some 14,000 tons of ore were offset by deliveries of 14,000 tons from mines to mills for no net gain at the end of the year in recoverable reserves.

The significance of this situation becomes clear when we consider that, if we are to meet the "reference" case projections for 1985 and 1990, annual discovery rates between 1976 and 1982 will have to average 1½ times the best historical performance year of 1969-1970. To meet the "expanded case" projections, average find rates will have to double the 1969-70 discovery rate.

The Forum's Mining and Milling Committee has indicated that it is unrealistic to believe that exploration can be adequately expanded under the present industry condition of low profitability and lack of forward commitment by the utility industry at prices which will provide a reasonable return on exploration investment. In this connection, the Mining & Milling Committee estimates that to meet the "reference case" rate of growth, it will be necessary for the uranium industry to invest in exploration and other capital expenditures the equivalent of approximately $5.40 per pound of uranium concentrates delivered during the next seven years. To meet the "expanded case" projections, they estimate that this investment commitment will increase to approximately $8 per pound.

In commenting on the likelihood that providing additional supplies of uranium will lead to higher costs, it should be pointed out that such costs are unlikely to make a significant difference in a comparison between nuclear and fossil fuels. For example, an increase from $8 to $12 per pound of yellowcake for an LWR [light-water reactor] reactor operating with plutonium recycle would increase power costs about 0.25 mil per kwh, the equivalent of an increase in oil of about 14 cents per barrel.

Given the proper incentives, it would appear possible for the uranium industry to mount an exploration effort 1½ times that of 1969-70, starting in 1974; and that by increasing that rate by 15-20 percent per year for the next 8-10 years, it would be possible to sustain a discovery rate of approximately 60,000 tons of U_3O_8 per year.

As to what steps might be taken to improve the uranium supply-demand situation, a number of suggestions have been made.

One is that utility managements should be encouraged to make commitments for their uranium supply as far in advance of their actual needs as their licensing schedules permit. One specific suggestion is that utilities holding construction permits should be obligated to demonstrate "procurement responsibility" by having in hand contracts for the uranium which will be required to meet their fuel needs for the initial 5-10 years of reactor operation. The thrust of this suggestion may have been indirectly satisfied in part by the AEC's new enrichment service criteria and contracting arrangements. Although the AEC has not spelled out the provisions of the enrichment contracts which will specify lead times for delivery of feed material, the longer lead times associated with commitments for enrichment services should provide incentive for the uranium industry to increase its exploration efforts.

A factor which is believed to work against both the investment of exploration capital and utility forward commitments to purchase uranium is the uncertainties associated with the embargo on the importation of uranium for domestic use. Both the uranium industry and the utility industry seemed agreed on the need for a viable domestic uranium supply capability, as well as on the need for establishing and publishing at an early date the criteria that will be used in handling the embargo question in the future.

It is of interest to compare the cost of importing uranium, if this should prove necessary or desirable, with the cost of importing oil. With uranium yellowcake at $8 per pound, oil would have to be imported at 15 to 28 cents per barrel to give the same direct fuel cost for power generation in LWR reactors. (This range depends on several variables including plutonium recycle.) Estimates of the cost of important oil in the 1980s range from $4.50 to about $8 per barrel or roughly 30 times as much.

Two other incentives identified by the uranium industry for assuring a continuing supply of domestic uranium are a continuation of a favorable tax environmental and timely access to public lands, both of which will tend to encourage exploration and development.

URANIUM ENRICHMENT

The report of the Forum's ad hoc Senior Management Uranium Enrichment Policy Committee, published last October, estimated that between the end of 1982 and the end of 1985, it would be necessary to put into operation as much new enrichment capacity as the AEC will have available in its three gaseous diffusion plants, once the combined capacity of those plants has been increased by the CIP [Cascade Improvement Program] and CUP [Cascade Uprating Program] programs. This is equivalent to four new plants, each with an annual separative work capability of approximately 7,000 MTSWU [Metric Tonnes of Separative Work Units]. If one extends the "reference case" rate of growth to 1990, the annual demand for separative work will have increased to 53,100 MTSWU and the cumulative demand will have nearly doubled in the 1985-90 period. If that demand is to be met, it will necessitate putting into operation five more 7,000 MTSWU plants in the same 1985-90 period; i.e., a total of nine plants over and above the three AEC-owned plants operating today. This assumes that in this period we will continue to meet the separative work requirements of overseas customers to the extent that they are not capable of meeting them with their own enrichment facilities. The "expanded case" rate of growth, which of course calls for an expansion of domestic nuclear power growth only, would require putting into operation three more plants in the 1985-90 period or a total of 12 plants between now and the end of 1990.

In setting forth these requirements for new separative work capacity, we have assumed a tails assay of 0.275 and 100 percent plutonium recycle in the LWR's in which it will be generated. If for any reason—e.g., licensing prohibitions or use of the plutonium for fast breeder inventories—full recycle does not prove to be the case, the requirements for separative work will be approximately 15 percent higher than we have indicated above.

In light of the number of new enrichment plants required for both the "reference" and "expanded" cases, it is extremely important that attention be focused on two prerequisities: capital investment requirements and lead times.

Irrespective of the technology selected, the capital investment requirements are expected to approximate $2 billion per 7,000 MTSWU plant, taking into account the capital requirements of the supporting electric power plant.

Because it will take 8-10 years to bring a new enrichment plant into operation, once the commitment to build the plant has been made, decisions have to be made and work initiated far in advance of the actual need for the new capacity. Although somewhat different considerations underlie the financing, construction and operation of an enrichment plant by the AEC in contrast to the way in which these responsibilities would be handled by the private sector, there is no reason to believe that the lead times in the two instances would be significantly different.

As the earlier Forum enrichment report pointed out, the AEC should—taking into account the full range of risks and unknowns involved in such a first-of-a-kind venture—formulate a set of guidelines for industrial involvement addressed to the following considerations, among others:

- Requisite policies and procedures for transferring AEC technology to private industry in such a manner as to permit industry to assimilate and utilize the technology without delay.
- Licensing criteria relating to financial responsibility, antitrust matters, health and safety and environmental concerns.
- Disposition of existing domestic and overseas contracts.
- AEC's continued operation of the existing plants with respect to market allocation and pricing in competing with a private plant.

This Committee shares the views of an earlier Forum committee convened to review AEC's proposed amendments to 10 CFR 25 which the AEC has set forth as a mechanism, albeit not a very effective one, for industrial participation. The committee observed that unless a private party wishing to enter the uranium enrichment area is afforded sufficient freedom to raise the requisite capital, attract a market for its product, develop a proprietary position commensurate with an assurance of a reasonable return on its investment, and communicate freely on unclassified subject matter, the responsibility for building and operating enrichment facilities is not going to be assumed by the private sector.

In view of the circumstance that has prompted these hearings—namely, AEC's proposal to revise its uranium enrichment services criteria and contracting arrangements, I should point out that the Committee on Resource Needs has not had an opportunity to consider the implications of the proposed changes.

Management in the nuclear industry in both the supplier and utility sectors is becoming increasingly aware of the need for expanding enrichment capacity. Purchasers of enrichment services are quite understandably concerned about the tougher provisions of the new contract terms, particularly regarding lead times for purchase commitments, downpayments, and the financial penalties associated with cancellation. On the other hand, most of these purchasers realize that their long-term interests outweigh their short-term concerns and that if the contract revisions, together with recently indicated price increases, assure that added enrichment capacity will be available in the early 1980s, these are steps that should be taken. They also realize that they will have to accept comparable long-term contracts to induce the entry of private enterprise into the enrichment service area.

With respect to the price increase, there seems to be little doubt that the price for enrichment services from new industrial plants will be appreciably higher than the AEC's current price of $32 kg of separative work. If one were to assume, simply as a matter of speculation, a price of $48 per kg of separative work, this would be 50 percent higher than the current price and 33 percent higher than the new price announced by the AEC. If one assumes full plutonium recycle, the impact of the $48 price, including inventory costs, would amount to about 0.2 mill per kwh or the equivalent of about 12 cents per barrel of oil. While a 10 percent change in fuel cycle costs cannot be considered insignificant, there is little reason to believe that it would influence a comparison of nuclear and fossil power costs in any major way.

The one aspect of the AEC's proposed new enrichment criteria and contracting arrangements that I trust will be developed more fully during the course of these hearings is exactly how they will lead to the installation of new enrichment capacity and to the entry of the private sector into the enrichment area. Certainly, the steps proposed by the AEC seem to be in that direction, but the implementing mechanism and procedures are far from clear.

TECHNICAL AND CONSTRUCTION MANPOWER

The Committee feels that an adequate supply of engineers and construction craft labor may indeed be a very serious problem but we cannot specifically identify it as a problem unique to nuclear power or even to the total electric power program inasmuch as even the maximum growth projections don't involve more than a few percent of the overall projected demand for engineers and construction craft labor. For example, our estimates show a requirement of some 25,000 technical personnel to support the engineering and construction effort that will be involved in the "reference case" and some 33,000 for the "expanded case." Added to these numbers will be some 83,000 and 125,000 operating and maintenance personnel, although not all technical, that will be required in supplying hardware, materials and services for the "reference" and "expanded" cases respectively.

We believe that the nuclear industry can cope with the problem of attracting, training and effectively utilizing the necessary manpower—if it is available.

The overall industrial engineering and construction activity in the 1980s can be expected to be very high as reflected by estimates of the capital investments required, especially from all aspects of energy supply and use including electric power and related services. The supply of engineers from the universities is currently on the decline, and while it may change, the "lead time" to obtain capable and experienced engineers imposes a serious problem.

The supply of craft construction labor is not growing at anything like the rate required. When one combines this fact with prevalent levels of productivity, a growing lack of mobility, high labor costs, and restrictive practices, it leads to a real concern as to whether a shortage of construction craft labor may not prove to be the most serious of our potential bottlenecks.

We must also note that current AEC licensing practices have led to a "consumption" of engineering man-hours in design and field supervision and to a "consumption" of construction man-hours far beyond earlier expectations and, to the best of our knowledge, far beyond the requirements of any other industry. We would hope that adoption of simplified licensing practices reducing duplication and unnecessary detail could lead to reduction of engineering man-hours, and that similar reductions could be made in the field supervision and craft labor.

Larger equipment, larger units, and multiple-unit plants have all been important not only in reducing costs but also in reducing manpower and material requirements in all phases of the nuclear industry. These trends are expected to and must continue if the requirements of an expanding nuclear power industry are to be met.

We believe that it is essential to seek to restore the normal output of engineers and scientists from our universities.

We believe it is essential to broaden the base for craft labor; i.e., increase the number of skilled construction workers, to find ways to increase productivity (partly by using less skilled labor where possible), and to restore the incentives for mobility so essential for building major new facilities in the "out-of-the-way" places which seem to be the destined locations for power plants.

FINANCING

The Committee has made rough projections of the financing requirements for nuclear power plants and related facilities, and these will appear in the final report. To give you some idea, however, of the "astronomical" sums involved, it has been estimated that to meet its projected demands for electric power, the utility industry will have to invest more than $650 billion in capital expenditures, in terms of 1980 dollars, between now and 1990. This is more than five times the investment the utility industry committed in the comparable time span of 1956-1972. If the industry should embark on a building program represented by the "expanded case," an additional $64 billion would be called for in plant and nuclear fuel expenditures.

Projected capital expenditure requirements in "related" facilities are much smaller, but not insignificant. It is estimated that some $27 billion in capital expenditures will be made by those supplying nuclear power plant equipment, materials and services in the 1973-1990 period to meet the needs of the "reference case." An additional $19 billion in capital expenditures would be necessary to meet the "expanded case" requirements.

While the proportions of the expected GNP which will have to be devoted to the "energy" industry and its construction requirements (investment) will increase substantially, and nuclear expansion will contribute significantly to these added requirements, the proportions required do not appear to be in any way impossible of achievement, even though the absolute dollar numbers at first appear alarming.

There undoubtedly will be an overall need to develop new mechanisms for large-scale financing and attracting investment capital. However, we believe this can be done if—and only if the industries and facilities being financed are viable financial investments. In short, the utilities, the manufacturers, and those providing services must have healthy and profitable businesses with adequate revenues, profits, and cash flows to attract and justify the investments required. While this may not sound very profound, it is of extreme importance and does, indeed, often seem to be overlooked in practice.

The viability of the overall electric power industry depends largely on the economic health of individual utilities as reflected by stock prices, bond ratings, and bond interest rates, and the willingness of regulatory agencies to permit adequate earnings levels on a timely basis. Many utilities have been financially hurt by lengthy delays in building and licensing new nuclear facilities.

Listed below are some possible steps which might provide incentives for investment in nuclear power projects:

- Strengthen existing legislation which discourages the flow-through of the tax benefits deriving from the investment tax credit and accelerated depreciation. The deferred taxes generated are an important contribution to internal cash generation and thus ease utilities' demands on the capital market.
- Increasing the investment tax credit selectively for projects such as nuclear power plants which result in a reduction in our domestic energy deficit and in a longer term benefit to the consumer. Additionally, consideration should be given to allow utilities to realize a portion of the investment tax credit benefits prior to the in-service date of the plant.
- The inclusion of plant construction costs in utilities' rate base as incurred.

The Committee has concluded that the problems of providing adequate manufacturing facilities, other materials and components, processing facilities, etc., for both the "reference" and "expanded" cases, can be taken care of in orderly and straightforward manner within the time period available, once the requirements become apparent—provided that prompt action is taken in the areas noted above. The Committee's rationale for this conclusion will be set forth in the final report.

Appendix J

Gas: One Company's Approach

Pacific Lighting Corporation, the parent company of Southern California Gas Company—the nation's largest in terms of end-use customers—provided detailed information on the problems it feels are most significant, and on its approaches to them.

Excerpts from the company's comments, prepared for this study in response to an outline, are included here as an example of the outlook of one of the largest and most sophisticated companies in the industry:

In the discussion that follows we [Pacific Lighting] have conformed with the major topics as presented in the outline, but have not tried to comment on each specific item under the major topic outline. Our approach was to explain our gas supply activities as they may apply to these various items. In the discussion we have included those sub-topics which we think are appropriate to our Company's operations.

EXPLORATION, DEVELOPMENT AND PRODUCTION

Almost all Pacific Lighting's various potential gas supply projects require some involvement in exploration, development and production. Specific supply projects that best reflect this involvement are discussed below.

Pacific Lighting/Transwestern Joint Venture

This joint venture was designed to supply gas to Transwestern's pipeline to counteract the decline in supply from other sources. Transwestern is the operator for the joint venture. The two companies agreed on a three-year budget for this operation. Within this budget, Transwestern works up the prospects, makes its evaluation, and, if acceptable to them, recommends the joint venture's participation. Pacific Lighting reviews Transwestern's recommendation and all the information which Transwestern used to make its recommendation, such as geological and geophysical data, cost estimates and agreements which would be necessary to drill for and produce any hydrocarbons discovered. Pacific Lighting then decides whether or not it wants to participate in the prospect. After this decision has been made, Pacific Lighting keeps informed of the progress of the exploration and development.

Australia/Palm Valley Prospect

Magellan Petroleum Company, through its exploration program, discovered what appears to be a major gas field in Palm Valley in the interior of Australia. The United States energy situation prompted Magellan to contact Pacific Lighting to determine our interest in developing the field for possible export to the west coast of the United States. Pacific Lighting reviewed all available geological, seismological and drilling data and found the prospect to be attractive. We employed Bechtel Corporation to make a preliminary estimate of the cost of pipelining the gas to the Australian coast, liquefying it and shipping the LNG to California. The cost determined by Bechtel appeared to be well within the range of the estimated cost of supplies from other areas. An agreement was then signed with Magellan to fund additional drilling in the area to delineate the size of the reservoir.

Pacific Lighting entered into the agreement contingent upon receiving approval of the California Public Utilities Commission to recover the cost of drilling through rates. It also made the agreement contingent upon obtaining an export permit from the Australian government for the LNG. The Company has not yet received the export license or the authorization from the CPUC to fund the program. A decision to drill the first delineation well was made in late 1972. The need for additional supplies and the relatively low cost of this first well supported the decision. If the drilling confirms the potential of the area, we may advance up to $4.4 million, the estimated cost of the delineation program, in return for a right to buy the gas. Pacific Lighting will purchase the gas from Magellan at a price set forth in the agreement. If sufficient reserves are proved up through this delineation program and the necessary approvals are obtained, we will advance up the $50 million for development well drilling.

This amount of drilling is estimated to develop sufficient reserves for a 500 million cubic feet per day project. We will recover our advance payments to Magellan from the proceeds of any hydrocarbon sales from the concession.

Most of Pacific Lighting's funding of exploration and development activities in Canada and Alaska generally follow the pattern of Palm Valley Australia. Advance payments to producers to explore for and develop gas is becoming the common method of obtaining a dedication of gas reserves. In exchange for these advance payments, Pacific Lighting gets a call on a certain percentage of the gas discovered and recovers its advance payments through the sale of the discovered gas.

We are also participating in ventures wherein we obtain a working interest in the gas discovered. These may develop into significant new sources of gas, but at this time, they are generally longer range projects than those which involve our making advance payments.

MARINE TRANSPORTATION

Pacific Lighting's interest in marine transportation started with our initial studies of the South Alaska LNG Project. Since it now appears that this will become an important new source of supply for Southern California, interest in this activity has increased sharply. To determine the economic size and speed of vessels which will be used in our various LNG projects, we funded a ship parametric study. On the basis of this study we are able to determine the trade-off between speed and size and are able to more logically determine the type vessel needed for our various LNG projects.

Since we first became interested in South Alaska as a source of LNG, our first shipyard contacts were with U.S. companies. It now appears that, even for our foreign LNG projects, construction in U.S. shipyards is preferable to foreign construction: (1) construction in foreign yards tends to worsen the country's balance of payments position and thereby hinders the obtaining of regulatory approval for such projects; (2) if a construction differential subsidy can be obtained for the construction of ships to be used in foreign LNG project, this removes the economic burden associated with U.S. ship construction; (3) financing of ships constructed in U.S. yards is generally more favorable than for those constructed in foreign yards.

Pacific Lighting, in its dealing with the shipyards, has followed the policy that it will not select a specific ship design for its LNG vessels. We still rely upon the decision of the shipyard selected to build the ship. We will, however, obtain an adequate builder guarantee and take measures to ensure that the vessel meets Coast Guard and American Bureau of Shipping standards.

The recent increase in LNG activity has reduced the availability of space in shipyards. Pacific Lighting, in order to minimize the time required to obtain a vessel, will shortly commit for the building of a ship, even though it has not received the necessary regulatory approvals for a project. The risk involved in such a decision is outweighed by our urgent need for gas, and our interest in LNG as a potential source of new supply.

Pacific Lighting has, for some time, been studying the desirability of owning, instead of chartering, LNG vessels. At this time, a decision has not yet been made, but it will be based in part on which method of financing yields the lowest transport cost to the consumer. The risk and reward to both stockholders and rate payers that attaches to the various forms of financing will also weigh heavily in the decision.

PIPELINE TRANSPORTATION

At the present time, Pacific Lighting is pursuing pipeline gas from Canada and Alaska. The Gas Arctic/Northwest Project Study Group, of which we are a member, is finalizing the engineering, environmental and other studies required for applications with the Canadian National Energy Board and the Federal Power Commission to seek authority to construct the Arctic Gas pipeline from the North Slope of Alaska to the Mackenzie River Delta to the lower Forty-eight States. Pacific Lighting has the option to acquire an equity interest in the pipeline which would be available to bring supplies South. If we are successful in obtaining the Far North reserves and the necessary permits to import Canadian gas, our share will flow through the proposed Arctic Gas pipeline to the U.S. border and thence to the West.

MARKETING

Pacific Lighting's supply acquisition policy is based upon a marketing policy to satisfy, at a minimum, our customers firm requirements and the requirements of those interruptible customers who cannot readily convert to alternate fuel on a year-round basis. In addition to this minimum level of service, we plan to market supplies to other interruptible customers. Service to this class would vary from time to time as increments reached the market place and became absorbed. Such service would ensure that a reasonable level of service could be maintained over time. The advantages of serving interruptible customers has been explained earlier, but it must be emphasized that this level of supply substantially benefits the firm customer as well as the interruptible customer. Since the firm customers have first call on all gas supply and none is dedicated to serve any other segment of our market, it is the position of Pacific Lighting that the relatively high-priced new increments of gas could be rolled in with the price of existing supplies and the resulting average cost of gas used in setting rates. Under the policy, gas can be priced competitively.

Because of the supply shortage, the Company's advertising program now emphasizes the conservation of energy. In addition to specific conservation steps designed to decrease the demand for gas, we point out the energy-saving benefits of gas-consuming appliances for heating, cooking, water heating and clothes drying over electrical appliances. No effort is made to promote added uses of gas.

RESEARCH AND DEVELOPMENT

Pacific Lighting sponsors several areas of research. It is participating in an industry fuel cell program which is designed to develop fuel cell power plants capable of producing economically attractive electricity while complying with the highest environmental and conservation standards. Pacific Lighting is also continuing its participation in various research programs designed to improve the technology of converting coal to pipeline quality gas. Pacific Lighting's location near large deposits of western coal makes this endeavor especially worthwhile.

CONSERVATION

Pacific Lighting's current advertising program is designed to promote those uses of gas which conserve energy. Conservation of energy will also be considered in Pacific Lighting's evaluation of LNG projects, methanol projects, and others which generally involve the use of fuel in a conversion process. An example of this would be Pacific Lighting's work in developing a design for reliquefaction machinery to be placed aboard ship so that the natural boil-off from the LNG tanks can be reliquefied rather than be burned in the ship's boilers.

DEFINING CORPORATE OBLIGATIONS

Pacific Lighting is dedicated to carrying out all its utility obligations and to the policy of obtaining sufficient gas to supply the needs of its firm customers and high priority industrial customers. While carrying out this policy, it also accepts its social responsibility to protect the environment.

Pacific lighting is carrying out its obligation to its shareholders by keeping its involvement in gas supply matters generally in accord with its historical utility operations.

In 1973, the California Public Utilities Commission approved an application which allows us to increase our rates specifically to recover the costs of Commission-approved exploration and development programs. Benefits from these programs would be passed on to the rate payers. Stockholders are relieved of the risk involved in acquiring large volumes of additional gas and, as a result, receive no return on these activities.

RECOMMENDATIONS ON PUBLIC POLICY

De-Regulation of Gas

We see no panacea in de-regulation of gas. We believed that regulation could respond by establishing new gas prices at something approaching a market clearing level in order to stimulate exploration and drilling. Since this so-called optional pricing procedure has run into legal obstacles which have impaired its effectiveness in stimulating new supplies, we now support congressional action to de-regulate new gas supplies.

Relaxing price controls for *old* gas would bring no new gas to the market, but would impose severe penalties on existing customers. For this reason, we see no benefit in relaxing of price controls for old gas.

Centralizing Government Decision-making for Energy

The present arrangement of multiple departments and agencies in government with their divided responsibilities and authorities, causes considerable confusion and delays to the energy industry in trying to meet the nation's energy needs. We believe a single agency should be formed which would consolidate the existing scattered agencies and departments in government which deal with energy. Further, we believe this agency should have its authority and responsibilities clearly defined. Among this department's more important responsibilities would be the development of a national energy policy, the regulation of those energy activities clearly authorized by legislation, and the administration of government energy programs.

We believe this new Energy Department should report to the Executive branch of government—cabinet level—and that formation of a Council of National Energy (much as the Council on Environmental Quality) would also be desirable.

National Security

According to all of the recent authoritative forecasts of U.S. energy demand and supply, the nation is now, and will become increasingly dependent on imported energy. The U.S. economy is the most industrialized in the world, and it is totally dependent on energy to function properly. We believe that from a broad, national point of view, the matter of national security in energy supply continuity (or the insurance against energy supply interruption) needs to be analyzed and evaluated. The ultimate objective would be the development of guidelines for a secure energy supply for the nation's energy industry.

We see these guidelines balancing the needs of keeping the nation's energy industry and its contribution to the overall economy health, while recognizing the benefits from utilizing readily available and inexpensive foreign energy sources to fill the nation's energy deficiency. Further, each barrel of oil imported helps to conserve the nation's dwindling energy pantry. We see security for the nation while using imported energy sources as coming from multiple supply sources, emergency supply stockpiles in a ready-to-use state, and reserves earmarked for quick development but held off the market.

Environmental Protection Considerations

We believe that protection of environmental quality is important, and we support the nation's goals for clean air, water and attractive surrounding. We do question whether the EPA and government bodies have approached the nation's environmental goals in a reasoned manner with regard to energy supply. We believe certain changes are in order, with the following specific suggestions:

A. A more reasoned approach to environmental quality which recognizes that any form of life intrudes on other forms of life, and, especially, that all social animals, including humans, alter their environment somewhat to better survive. The barriers to energy supply have become extreme in some cases and all out of balance when benefits are weighed against costs. The Alaska pipeline is a case in point; and progress should not be blocked merely because there will be a change in the environment.

B. Government indemnification for penalty costs incurred by industry, who in good faith build new facilities according to known environmental regulations and find that the facilities are inadequate or outmoded by later environmental regulation changes.

   With this approach, we believe refining capacity for desulfurization of oil, and the processing plants for desulfurization of coal, would be built sooner. The energy industry is very reluctant to construct facilities costing over $100 million each when their utility is doubtful. For example, tighter controls which outmode the plant, or relaxed controls making the plant unnecessary.

C. Avoidance of blanket, year-round tight fuel quality limits when the period of actual need is only part of the year. That is, restrict the use of clean fuels to those areas and periods of time where its use will be of benefit. We believe this would conserve clean fuels in tight supply and allow use of more plentiful fuels.

D. Establishing air and water quality standards based primarily on known technical data; not arbitrarily tight limits. If an energy industry standards-setting organization (such as ANSI) believed, through actions of its members, that some air or water quality standard was unnecessarily restrictive, it should have the right to challenge the regulation. This challenge would require substantiation with supportive data from the industry for a "more appropriate" standard. The regulatory body would be granted 12 months to develop technical data to support the regulation in question. If support was not developed, the regulation would be required to revert to the proposed "more appropriate" standard.

E. In particular areas or circumstances where a region desired to establish environmental standards tighter than those supportable by technical data, it may establish such standards only with the provision that the general populace will bear the added costs incurred, and the standards are physically possible.

Impact of Jones Act

We believe that the Jones Act results in inequities for the overseas shipment of oil or LNG from Alaska to the contiguous 48 states by requiring U.S. bottoms and crews. Foreign bottoms and crews are substantially less expensive than U.S., but the Jones Act bars them from the Alaska-U.S. trade.

In foreign trade including the Far East, mid East or Carribean oil-producing areas, foreign bottoms and crews can be used to import oil or LNG to the U.S. To achieve parity for U.S. shipping in foreign trade, the government provides subsidies. We believe that either the Jones Act should be revoked or the maritime subsidy program should be revised to provide overseas energy shipping from Alaska with the same benefits offered U.S. overseas shipping of energy from foreign sources.

End Use Controls

We recognize that the nation is currently in a deficient supply situation for environmentally clean fuels. As a result, various means of conserving or allocating the available clean fuel supplies are being considered by government. We are opposed to the arbitrary application of end use controls without regard for regional air quality needs, capability to use alternate fuels and availability of alternate fuels. We also oppose any measures that inhibit, rather than contribute to, solutions of the nation's energy problems—for any energy source: oil, coal and nuclear as well as gas.

We further believe that where we and our customers sustain economic burdens or cost penalties as a result of allocation or end use controls which deprive us of gas we have under contract, that appropriate compensation is due from those who benefited by the action.

Government Energy Business

The government energy businesses and related activities such as TVA, BOA, REA, Cooperative Power Associations and Municipal Power and Gas Companies were all started as limited endeavors to provide necessary services under government sponsorship—which was considered the only means for the services to come into being. All of these enterprises have both matured and grown. They are now viable enterprises, well able to survive without continued government support. We believe that these government energy businesses should issue stock in the manner of COMSAT, with the initial stock ownership shared between government and the private sector.

Ideally, we believe stock ownership could be completely held outside of government, with these enterprises taking a full role as taxpayers supporting the government directly in all areas, in lieu of their present indirect contributions to limited areas.

Leasing of Federal Controlled Lands

Leasing of federally owned or controlled lands should be expanded at a rate consistent with national needs. If it is argued that these areas should be preserved (if not entirely, at least well into the future), then the nation's energy deficiency will have to be made up by larger oil imports. If, on the other hand, it's desired to hold oil imports down, then leasing must be expanded to add to the nation's energy supplies. Further, the nation needs to develop technology for deep drilling and bottom hole production; and this development isn't likely to occur without the encouragement of new drilling possibilities from an expanded leasing policy.

Problems with Governmental Delays

We believe that delays in governmental processes regarding the energy industry must be kept to a minimum. With labor and material costs escalating at a rapid rate, the regulated energy industries must be able to obtain corresponding rate relief or suffer erosion of allowed earnings with a consequent adverse impact on their ability to raise capital. In addition, entirely new cost burdens from government safety, health and environmental quality regulations must be reckoned with. We believe the normal hearing procedures for obtaining rate relief for these new cost burdens (rapid wage and material escalations and new government regulations) with are totally outside management control, are too slow and need to be replaced.

We propose that rules be established which allow direct flow-through into rates of all major costs not subject to direct management control, much as the so-called purchased gas adjustment clauses currently in effect in some locations. This would not cover those areas where management does have control—such as the establishing of new fuel purchase contract. Those should continue to receive normal hearing procedures. Once the new contract has been heard and approved, the subsequent implementation of that contract through secondary purchasers or established time changes, should be flowed through.

We believe any unnecessary delays in governmental decisions will be particularly harmful to the energy industry and the nation. In the first case, the nation will need to increase its inflow of foreign energy over the coming years. Other nations also will be competing for energy supplies during this time, and the ability to consummate negotiations quickly is important. In some cases, arms-length negotiations are impossible to conduct when a negotiator must qualify his position subject to governmental approval, with no time limit on the approval date. Such situations invariably incur cost premiums and may result in the loss of new energy supply.

In the second case, the nation can be harmed irreparably by government delays on energy industry decisions. The nation's economy runs on energy, and any energy deficiency impairs not just a portion of the energy industry, but the whole economy. Further, opportunities once lost cannot be made up later by the industry—other nations will have established their positions in any void created by U.S. delays. We are not looking at an internal situation where the resources are at our command and will remain there for our use whenever we wish to take them. With current oil imports expected to reach over half of U.S. needs (and world demand growing rapidly), timely decisions are imperative.

Appendix K

The Potential of Coal

In December of 1973 the National Coal Association was asked by the Administrator of the Federal Energy Office to outline the degree to which the bituminous coal industry could expand production to help meet the energy crisis. On December 20, 1973 representatives of the Board of Directors of the National Coal Association presented a white paper entitled "The Potential of Coal to Meet the Energy Crisis" to Mr. Simon, then Administrator. The paper presents the views of the industry on both short and long-term steps necessary to substantially increase coal production. It is presented in its entirety following:

INTRODUCTION

The coal industry will do its utmost to expand to meet the energy crisis. We accept the responsibility for providing a major share of the national energy supply.

But we would be remiss in our duty to the American people if we did not point out that short-term expansion is inevitably and closely intertwined with long-term development. The American people recognize the critical status of energy. We point out that we must develop the coal resource of the United States if the present crisis is not to develop into a future catastrophe. Thus it is incumbent upon those who develop national policy to work with all interested segments of the coal industry to fashion an expansionary coal policy designed to insure the national goal of energy self-sufficiency in the next decade.

This is an attainable national goal if coal can make the necessary investment now in coal production capacity and if the nation is willing to construct the type of institutional framework favorable to the rapid development of coal.

CURRENT CAPACITY

Projected 1973 coal production is 590 million tons. This represents a decline from 595 million tons produced in 1972, and 603 million tons mined in 1970.

More significant, perhaps, is the fact that coal demand in 1973 was projected to be 609 million tons. This estimate was made prior to the Arab oil embargo; actual demand is probably more.

Coal mining capacity has remained stagnant for several years in the face of constantly increasing demands. For example, a recent study published by the United States Bureau of Mines estimates 1972 production potential at 741 million tons; 1971 at 736 million tons; and 1970 at 740 million tons. These estimates are based on a *280 day working year,* a figure which has never been attained in the history of the industry. More realistically, at a 225 working day level, which is relatively common in the industry, and is the number of days worked last year, the estimates of potential capacity would be 595 million tons in 1972; 591 million tons in 1971 and 594 million tons in 1970.

Whatever the relative numbers, however, the obvious truth of stagnation in production capacity is startling.

Production capacity, however, is not static. It can be circumscribed by a host of inhibiting factors, or it can be expanded if the external and internal variables surrounding production can be arranged to encourage it.

Recently the National Coal Association completed a survey to determine how far production capacity could be expanded. In brief, we found the potential for increased production in the following orders of magnitude given present machinery and labor in the bituminous coal industry:

Tons Per Week	
441,000	30 days
913,000	3 months
1,355,000	6 months
2,087,000	1 year

These additional tonnages could make available 50 million tons or more of coal during the next 12 months. They represent the willingness of the coal industry to help to meet our current energy crisis. Implicit in them, however, are the assumptions that the variables which now inhibit production would be changed to encourage it. These assumptions include labor stability; adequate supplies and materials; no inhibiting price restrictions; departure from optimal mining plans when needed to meet short-term production requirements; stringent but non-harassing enforcement by MESA of the Coal Mine Health and Safety Act of 1969; relaxation, within health requirements, of air pollution controls; proper regulation of surface mining with adequate reclamation but no direct or indirect prohibition.

Implicit also is the assumption that the crisis of the short term will evolve into a long-term national commitment to coal. Such a commitment made within the framework for an expansionary governmental policy and investment climate will do much to assure national energy self-sufficiency. Without such a national determination, not only the future capability of the coal industry but also its present expansibility is seriously in doubt.

THE INGREDIENTS OF GROWTH

The Commitment to Development

Clearly we are now at a turning point in our energy production and use patterns. The fallacy that we could safely depend on foreign supplies is now glaringly exposed. So too is our national propensity to waste, a propensity nurtured over the years by seemingly endless abundance.

The alternatives are starkly clear. We as a nation can develop our energy patrimony and use it with uncharacteristic wisdom, or we can drift from crisis to crisis until inevitably we will have to accept both a sharply lower standard of living and a position of dependency in international affairs.

If we choose the logical course, i.e., indigenous development, there must be two essential governmental inputs:

First, government must make and enforce a national commitment to indigenous energy resource development, particularly of coal. It must be as intense as our national commitment to full employment in 1946, to atomic energy development in 1954, and to the space program of the early 1960s. The commitment must not waiver even in the face of temporary changes in energy supply/demand conditions which make possible economic advantages through the use of foreign energy resources.

Second, there must be a national determination to conserve scarce fuels by substituting abundant fuels where possible. The use of coal instead of oil and natural gas for boiler fuel is a good case in point. Conversions of electric utility boilers to coal, now being undertaken, must be made permanent so that available petroleum will go to those uses for which it is uniquely suited.

Labor Stability

One of the most significant factors currently inhibiting coal production is labor. For the longer term the problem of manpower presents a major constraint to coal expansion.

Labor policies conducive to growth can be viewed in two ways: (1) Those things which inhibit the full use of today's productive capability. (2) Those things which in both the short and long term circumscribe growth.

In the category of limits to production fall such factors as wildcat strikes and high rates of absenteeism. Also involved is the difficulty which coal is now experiencing in securing front-line foremen and competent engineering personnel.

The limits to growth include such factors as the need for and scope of manpower development, the way in which coal mining as a career is viewed by the public at large as well as by those who tend to guide the career choices of young Americans, and the degree of cooperation among management, labor and government in fostering an industrial relations policy conducive to expansion.

Compounding the labor dilemma is the peculiar demographic makeup of today's work force in coal. The age distribution is bi-polar with peaks in the over-45 and the under-30 age groups, and a gap in the 30 to 45 years category. Thus, the industry is faced with a grandfather/grandson work force. It has relatively few of the experienced men 30 to 45 years old who stabilize a work force in other industries.

The industry must look to an increasingly younger work force at all levels and to some means of substituting training and development for years of experience. Whether this can be done is one of the most crucial issues affecting coal development on the scale necessary to make the nation self-sufficient in energy.

The coal industry is meeting its responsibility to pay its miners well; they are the highest paid workers in basic industry. They are skilled men who do a hard job with pride. If the coal industry is to meet its new challenge it must have their cooperation.

Financial Incentives

The coal industry's projected financial requirements are simply staggering. The National Petroleum Council's report a year ago on the assumption that Mid-East oil would continue to be available, estimated coal's capital needs to be $10-$15 billion by 1985. This is projected in 1970 dollars. For an industry with a current capitalization of $4 billion the magnitude of the task seems almost unattainable. However, such pessimism is unwarranted if certain things happen.

First, coal must compete for its investment funds. To do so successfully it must be an attractive investment opportunity with a competitive short- and long-range rate of return. Currently the industry simply does not have such a rate of return and thus the potential for development remains only that—a potential.

Second, current price restrictions which inhibit coal development must be removed. New mining capacity is badly needed, and yet it cannot be added without a more favorable return of investment. The Cost of Living Council gave credence to this factor in exempting long-term utility coal contracts, but the action was not nearly enough. A more proper course would be to remove coal from price restrictions entirely, at least until significant new coal capacity had been added and coal profit margins move upward to competitive levels.

Third, the heavy burden of "retroactivity"—benefits to miners who incurred the disease before the law required the industry to pay for benefits imposed by the black lung law—must be removed. If the industry must pay the potential billions of dollars for "retroactive" black lung disease, its ability to expand will be greatly impaired. This is especially true in the "traditional" mining segment of the industry, particularly those companies with a long history of underground mining in the Appalachian region.

Fourth, the tax structure of coal must be shifted to encourage expansion. NCA has proposed several major changes in the tax code to do this. Each of these individually and all of them collectively will enhance coal's financial attractiveness and should be considered by the Congress.

Long-Term Assurances

Coal mining is a long-term proposition. Today more than 70 percent of coal moving to utilities is under some form of long-term contract. New mines are usually opened only in conjunction with a long-term purchase agreement by a utility or other large coal consumer. Any program of expansion must be for the long-term, requiring contracts or other assurances that the product will have a market. Basic to the validity of such long-term arrangements are several factors.

First, there must be some willingness to compromise non-health related air pollution standards. To date, very little if any high-sulfur coal is under development and segments currently producing such coal are facing shutdown because of unreasonable sulfur-emission regulations. Few investors will risk capital for mine development when pollution requirements could make mines environmentally obsolete long before their normal pay-out period. Some mechanism is needed to permit the use of high-sulfur coal where possible, while developing and installing the technology for sulfur control. In this way the coal industry will have the necessary assurances that consumers can accept coal as a viable long-term energy resource.

Second, consumers, particularly utility and large industrial users, must know that they cannot expect to reconvert to old fuel consumption patterns once the current crisis is over. To attain the national goal of self-sufficiency there must be a national policy of coal use for boiler fuel purposes, as well as accelerated development of both gasification and liquefaction.

Third, legal inhibitions to expanded coal production must be both reasonable and predictable. Principal among these are leasing policies, the availability of water rights for coal development and the practicality of surface mining reclamation requirements. For example, the current status of surface mining legislation before the Congress has inhibited the development of surface mined coal throughout the U.S., especially in the West.

Adequate Supplies of Material

Coal, like any other industry, faces critical supply shortages. Lack of such items as diesel oil, lubricants, explosives and roof-bolts inhibit coal expansion. NCA has called this matter to the attention of government officials on numerous occasions. They have willingly helped us avoid serious disruption in supplies. But close attention must continue to be paid to this area so that prompt action can be taken to insure the flow of materials necessary for coal production.

Legislation

We have mentioned legislation briefly in previous paragraphs, but the legal framework surrounding coal is so crucial that more detail is justified.

1. **Surface Mining Legislation**: The surface mine legislation now in the House Interior Committee, or the bill passed by the Senate, is a disaster. Typical of the regressive natures of both bills is the Mansfield amendment, which would bar the development of more than 37 billion tons of western coal, and the Seiberling amendment, which would place coal at a competitive disadvantage. It makes no sense to disrupt 50 percent of our current coal production, especially at a time of critical need. Prohibitory legislation is not even remotely necessary now for sound reclamation can be carried out within normal mining processes.

2. **Air Pollution Regulations**: Some modifications of present air pollution standards are being incorporated in pending emergency energy legislation. We hope this action will be followed by a more permanent approach to reasonable national pollution abatement strategy geared to increasing energy availability. NCA believes that this can be done; we have outlined a program to do this, based upon:

- A stretch-out of time for compliance.
- A review of the mechanism for the establishment of standards.
- Alternative abatement strategies.
- Accelerated development and use of pollution control technology.
- The linking of abatement needs to specific points rather than broad geographic areas.

3. **Research and Development**: R&D means much to coal's future. The industry and its consumers must look for the type of advanced technology which will permit the greatly expanded production, distribution and consumption of coal within tolerable environmental conditions.

The National Coal Association has consistently supported the development of a sound national research and development program. We believe that such a program must include a coal research effort at least on a parity with nuclear research and a conceptual direction of the total energy R&D effort within the federal government. To this end we have consistently supported, at least in concept, the Energy Research and Development Administration, which to us would mean a centralization of research direction and a vastly expanded funding level for such programs.

One of the forgotten research priorities is the technology of the actual mining of coal. Extensive research programs on gasification and liquefaction, and advanced power systems will be useless if the coal needed for these technologies cannot be mined with present technology. Along with a change in the institutional structure for research and development, there must be a shift of emphasis so that the extraction of coal will be a major research and development priority.

However, research and development alone will not mine more coal. Research and development is but one subsystem of the total system of coal expansion. A great increase in coal production will only be possible if R&D is carried on in a legal and economic climate conducive to growth. It will require a national commitment by government, producers and consumers to use coal to the maximum degree possible.

4. The Coal Conversion Act of 1973: The National Coal Association supports the passage of S. 2652, the Coal Conversion Act of 1973. This Act represents the first step in the development of a national mandate to use the vast potential of our coal resources. We believe that this Act can be expanded to set forth the mandate for coal development explicitly and we have urged the Senate Interior Committee to do so. We hope that the Administration will support this action and use all of its influence to insure that the Coal Conversion Act of 1973 will be enacted and signed into law.

EXPORTS

Increased interest has been shown in the past several weeks on fuel exports. Some have suggested that since there is a shortage of coal in the United States, coal exports should be embargoed. This ignores several basic characteristics of the coal export market.

First, coal exports are essentially metallurgical grade coal. This coal is used primarily in the steel industries of western Europe, Canada, and Japan. The one notable exception is steam coal exported to Canada for use by Ontario Hydro. This coal often returns in the form of electric power; that section of the Canadian utility industry is intertied with U.S. utilities and serves as a valuable source of additional power for large portions of the northeastern part of the United States. The U.S. gets oil and gas from Canada and is in a poor position to initiate an embargo.

Second, most coal moving abroad goes under long-term contracts, many of which were signed several years ago. These contracts represent a good faith agreement between a U.S. producer and a foreign consumer on coal supply. The initiative for much of the export activity originally came from the federal government which looked upon coal as a major prop under our sagging balance of payments.

Third, some of the coal moving to foreign consumers comes from mines in which such consumers have substantial investment. Canadian, Japanese, French, German, and Italian consumers, for example, have invested many millions of dollars in coal production facilities, especially in the Appalachian area.

If the U.S. were to embargo coal exports it would disrupt the economies of many of our major allies. This is particularly true in Japan, which depends upon U.S. coal for its steel industry. Japan already faces staggering difficulties because of the Mid-East oil embargo.

TRANSPORTATION

Transportation bottlenecks must be broken if the use of coal is to expand as the nation requires. Consumers must depend in large part upon rail transport for large scale movements of coal. This matter is particularly critical in the Northeast, where the rail system is in virtual collapse because of bankruptcy proceedings and the railroads cannot attract capital in order to develop additional coal carrying capacity.

This is not the place to suggest a solution for the rail or transportation crisis. The coal industry stands to work with government and with the transportation industry to do what is necessary to insure coal moves from mine to market.

CONCLUSION

We are now at a major turning point in our history insofar as energy is concerned. There are only two alternatives open to us.

We can continue the policies of the past—policies which have brought us to our current impasse, and which inevitably must lead to national catastrophe. Inherent in such policies are the assumptions of continued energy abundance, of security of foreign supplies, and the wasteful production and use patterns traditional in our society.

Or, we can use the current crisis as a catalyst to action, as a prod to build an energy structure truly responsive to the nation's needs. That structure will be firmly rooted in the concepts of energy self-sufficiency through the development of our huge coal resources and use patterns which reflect the need for careful husbanding of our energy resources.

Index

About the Author

John E. Gray has been since 1972 an energy consultant to the U.S. Electric Utility Industry, the U.S. Federal Energy Administration, the Japanese Utility Industry Enrichment-Reprocessing Group, the MIT Center for Policy Alternatives and the Ford Foundation. He is a member of the Atlantic Council's Energy Committee and the American Institute of Chemical Engineers' Energy Committee. He is also a director of the Institute for Public Transportation of New York.

Mr. Gray was the founding president and chief executive of NUS Corporation, an energy supply/environmental consulting engineering firm, from 1960 to 1972. He previously held management and technical positions with the Duquesne Light Company, the U.S. Atomic Energy Commission, the U.S. Navy, General Electric Company and Westinghouse. He received a B.S. degree in Chemical Engineering from the University of Rhode Island in 1943.

Mr. Gray is co-author of Energy Research and Development (1974), and Nuclear Power (1973)–in Marks Standard Handbook for Mechanical Engineers, and editor of Shippingport [Atomic Power Station] Operations (1960.)